ADVANCED CIRCUITS AND SYSTEMS FOR HEALTHCARE AND SECURITY APPLICATIONS

VLSI devices downscaling is a very significant part of the design to improve the performance of VLSI industry outcomes, which results in high speed and low power of operation of integrated devices. The increasing use of VLSI circuits dealing with highly sensitive information, such as healthcare information, means adequate security measures are required to be taken for the secure storage and transmission. *Advanced Circuits and Systems for Healthcare and Security Applications* provides broader coverage of the basic aspects of advanced circuits and security and introduces the corresponding principles. By the end of this book, you will be familiarized with the theoretical frameworks, technical methodologies, and empirical research findings in the field to protect your computers and information from adversaries. Advanced circuits and the comprehensive material of this book will keep you interested and involved throughout.

This book is an integrated source that aims at understanding the basic concepts associated with the security of the advanced circuits and the cyber world as a first step towards achieving high-end protection from adversaries and hackers. The content includes theoretical frameworks and recent empirical findings in the field to understand the associated principles, key challenges, and recent real-time applications of the advanced circuits and cybersecurity. It illustrates the notions, models, and terminologies that are widely used in the area of circuits and security, identifies the existing security issues in the field, and evaluates the underlying factors that influence the security of the systems. It emphasizes the idea of understanding the motivation of the attackers to establish adequate security measures and to mitigate security attacks in a better way. This book also outlines the exciting areas of future research where the already-existing methodologies can be implemented. Moreover, this book is suitable for students, researchers, and professionals who are looking forward to carrying out research in the field of advanced circuits and systems for healthcare and security applications; faculty members across universities; and software developers.

ADVANCED CIRCUITS AND SYSTEMS FOR HEALTHCARE AND SECURITY APPLICATIONS

Edited by
Balwinder Raj, Brij B. Gupta,
and Jeetendra Singh

CRC Press
Taylor & Francis Group
Boca Raton London New York

CRC Press is an imprint of the
Taylor & Francis Group, an **informa** business

First edition published 2023
by CRC Press
6000 Broken Sound Parkway NW, Suite 300, Boca Raton, FL 33487-2742

and by CRC Press
4 Park Square, Milton Park, Abingdon, Oxon, OX14 4RN

CRC Press is an imprint of Taylor & Francis Group, LLC

© 2023 selection and editorial matter, Balwinder Raj, Brij B. Gupta and Jeetendra Singh; individual chapters, the contributors

Reasonable efforts have been made to publish reliable data and information, but the author and publisher cannot assume responsibility for the validity of all materials or the consequences of their use. The authors and publishers have attempted to trace the copyright holders of all material reproduced in this publication and apologize to copyright holders if permission to publish in this form has not been obtained. If any copyright material has not been acknowledged please write and let us know so we may rectify in any future reprint.

Except as permitted under U.S. Copyright Law, no part of this book may be reprinted, reproduced, transmitted, or utilized in any form by any electronic, mechanical, or other means, now known or hereafter invented, including photocopying, microfilming, and recording, or in any information storage or retrieval system, without written permission from the publishers.

For permission to photocopy or use material electronically from this work, access www.copyright.com or contact the Copyright Clearance Center, Inc. (CCC), 222 Rosewood Drive, Danvers, MA 01923, 978-750-8400. For works that are not available on CCC please contact mpkbookspermissions@tandf.co.uk

Trademark notice: Product or corporate names may be trademarks or registered trademarks and are used only for identification and explanation without intent to infringe.

ISBN: 978-1-032-03907-7 (hbk)
ISBN: 978-1-032-03908-4 (pbk)
ISBN: 978-1-003-18963-3 (ebk)

DOI: 10.1201/9781003189633

Typeset in Times
by MPS Limited, Dehradun

Dedicated to my wife, **Dr. Madhu Bala,** for her constant support during the course of this book

Balwinder Raj

Dedicated to my wife, **Varsha Gupta,** for her constant support during the course of this book

B. B. Gupta

Dedicated to my wife, **Dr. Chandra Kanti Singh,** for her constant support during the course of this book

Jeetendra Singh

Contents

Acknowledgments	ix
Preface	xi
Editors	xv

1 Design and Analysis of Charge Plasma-Based SiGe Vertical TFET for Biosensing Applications   1
Shailendra Singh, Sanjeev Kumar Bhalla, Jeetendra Singh, Shilpi Gupta, Balwinder Raj, and N. K. Yadav

2 Borospherene-Based Molecular Junctions for Sensing Applications   19
Jupinder Kaur, Ravinder Kumar, and Ravinder Singh Sawhney

3 Nanowire FETs for Healthcare Applications   39
Deep Singh and Amandeep Singh

4 TFET-Based Sensor Design for Healthcare Applications   69
Tulika Chawla, Mamta Khosla, and Balwinder Raj

5 Modeling and Simulation Analysis of TFET-Based Devices for Biosensor Applications   105
M. Arun Kumar and Meenakshi Devi

6 Security-Based Genetic Algorithms for Health Care   123
Neeraj Kumar Rathore

7 Role of High-Performance VLSI in the Advancement of Healthcare Systems   147
Jeetendra Singh, Balwant Raj, and Monirujjaman Khan

8 Trust-Based Security Model for Adaptive Decision Making in VANETs   161
Gurjot Kaur and Deepti Kakkar

9 Security Attacks and Challenges of VANETs   181
Manojkumar B. Kokare and Deepti Kakkar

10 **Energy-Efficient Approximate Multipliers for ML-Based Disease Detection Systems** 203
Zainab Aizaz and Kavita Khare

11 **Cross-Domain Analysis of Social Data and the Effect of Valence Shifters** 227
Aarti and Raju Pal

Index 249

Acknowledgments

Many people have contributed greatly to *Advanced Circuits and Systems for Healthcare and Security Applications*. We, the editors, would like to acknowledge all of them for their valuable help and generous ideas in improving the quality of this book. With our feelings of gratitude, we would like to introduce them in turn. The first mention is the authors and reviewers of each chapter of this book. Without their outstanding expertise, constructive reviews, and devoted effort, this comprehensive book would become something without contents. The second mention is the CRC Press/Taylor & Francis Group staff, especially Gabriella Williams and her team for their constant encouragement, continuous assistance, and untiring support. Without their technical support, this book would not be completed. The third mention is the editor's family for being the source of continuous love, unconditional support, and prayers not only for this work, but throughout our lives. Last, but far from least, we express our heartfelt thanks to the Almighty for bestowing over us the courage to face the complexities of life and complete this work.

Balwinder Raj
B. B. Gupta
Jeetendra Singh

Preface

Nowadays, advanced circuits and systems have become essential and are increasingly being used for healthcare and security applications. It has almost become impossible to imagine a world where people can live without these advanced electronic circuits and systems. With the rapid development in the electronics business, a number of inventions have come into picture including laptops, tablets, palmtops, and so forth used for healthcare and security applications. These are being used by individuals of different age groups and in almost all business processes because of the significant rise in the productivity of the work done and overall efficiency.

Chapter 1: This chapter deals with charged and neutral biomolecule detection with the suggested charge plasma-based SiGe heterojunction vertical TFET for the underlap region is analyzed. This draft stands out to the readers by distinguishing itself from other sensing devices currently on the market by demonstrating its greatest sensitivity in detecting charge and neutral biomolecules. As a result, the proposed device demonstrates improved performance for cost-effective biomedical diagnosis instrument construction, as it achieves increased sensitivity through a moderate optimum selection of geometrical parameters such as cavity length and thickness at the tunneling junction interface with appropriate biasing conditions. The biasing voltage V_{gs} is held constant at 1.5 V, whereas V_{ds} varies between 0.5 and 1.5 V. After analyzing its variation to different cavity dimension parameters with dielectric constant to its electrical characteristics, it can be concluded that the proposed charge plasma-based SiGe vertical TFET has a low leakage device current and is a highly sensitive factor for the change and neutral biomolecules.

Chapter 2: Two borospherene-based molecular junctions are designed, first by simply suspending the fullerene between the two electrodes (no bridge) and then by attaching the fullerene via single bonds with both the electrodes (with bridge). The current-voltage curve, transmission spectra, transmission eigen states, transmission pathways, and differential conductance are determined for both molecular junctions, taking into consideration the density functional theory and non-equilibrium Green's function formula. Borospherene molecular junction with bridge demonstrates superior transport behavior with highest value of current, 164284.76 nA at 1 V. Non-bridged borospherene MJ on the other hand showcases a large rectification ratio (1.47), making it an ideal choice to be exploited in rectifier applications. The superior transport phenomenon in case of bridged-borospherene MJ is due to more coupling between the fullerene and electrodes, a greater number of forward transmission pathways, and enhanced delocalization of molecular orbitals with participation of both the pie and sigma bonds in transmission. Also, this device assays Kondo Tunneling phenomenon with differential conductance approaching $2G_0$. At equilibrium, the conductance is $2.15G_0$, thus remarking the lifting of the coulomb blockade and paving ways for exploring this single molecule device in SET (single electron transistor) applications.

Chapter 3: In this chapter, the structure of advanced MOSFET has been discussed in brief. The modeling of nanoscale devices introduced ballistic transport and quantum transport of charge carriers. This chapter explains the various applications of nanowire FETs for healthcare applications such as biosensing. This chapter reviews the applications of nanowire FETs for virus detection, biomarker detection, DNA detection, and their applications for the discovery of drugs; in addition, it also describes the modeling of nanowires.

Chapter 4: This chapter covers extensive study and analysis of various TFET-based biosensors. it is concluded that TFET-based biosensors are superior compared to traditional biosensors in terms of sensitivity and response time due to their different carrier injection mechanism i.e., band to band tunneling mechanism. Further TFET-based sensors overcome the issue of short channel effects and power consumption too, which were the limitation of MOSFET-based sensors. Further, improving the sensitivity of TFET-based sensors can be done by employing various engineering on the device such as hetero dielectric material engineering, short-gate technique, and pocket doping technique etc. Furthermore, to overcome the issue of fabrication, charge plasma technology is used for the TFET-based biosensors. In recent times, diagnosis of COVID-19 pneumonia is done by a chest CT scan test and RT-PCR test. Numerous suspected persons may not be verified and isolated in a timely manner due to overworked medical staff in epidemic areas, which is alarming. There is an urgent need for more reliable, quick-response, cost-effective, and widely accessible analytical equipment or diagnostic techniques. So, in this respect, biosensors are powerful tools to detect such viruses or bioterrorism threats. Gas sensors particularly developed for human breath analysis are a reliable option for a fast and accurate detection of illnesses compared to existing conventional diagnosis procedures such as slow and intrusive blood tests. Further monitoring of gases is also very essential, as over-concentration of a few gases can damage the nervous system. Highly sensitive gas sensors can be designed using TFET device and charge plasma technology. At the present time, the demand for low-power and extremely sensitive photo sensors is rising at a surprising rate in numerous domains. For biomedical fields, highly sensitive photodetectors are required, so there we can use hybrid TMD-based TFET-based photosensors. This chapter provides researchers a quick review of previous reported biosensors, gas sensors, and photosensors to now and also gives an idea how to create future generation TFET-based different sensors with improved performance that are low-cost, highly sensitive, highly selective, more precise, more reliable, portable, and have a low response time.

Chapter 5: Different biosensor devices have been investigated in this chapter after the careful comparison of these other TFET device structures and performance using simulation tools. Due to the band to band tunneling (BTBT) conduction mechanism, the TFET device overcomes the limitation and reduces the short channel effect. Hence, a TFET-based biosensor has developed as a suitable candidate for superior sensitivity and response time than FET-based biosensor. In this chapter, the structure of the device and the parameters are investigated to design the TFET sensor devices. Hence, it will be more beneficial to create novel TFET devices for biosensor applications. The performance results show that the dielectric constant becomes constant, the thickness of the cavity grows smaller, the more positively charged, and the greater the sensitivity of the TFET device. Furthermore, simulation results show that the TFET structures can be applied for ultra-sensitive and low-consumption biosensor devices.

Chapter 6: A security-based genetic algorithm serves as a vigorous adaptive approach to remedyexploration and optimization issues. It is based on the catechistic searching techniquewhich solves the problem quickly i.e. guarantee to give the best solution with theoptimal solution and security may be obtained. They are more robust than conventional algorithms. GA does not break easily, unlike in older AI systems. The general GA consists of the following components: Initial population, calculate fitness function, selection, crossover, mutation, and stopping criteria. The basic operators used in genetic algorithms are encoding, selection, crossover, and mutation. The security-based GA is useful for medical diagnosis, curing diseases, etc.

Chapter 7: VLSI technology has grown in tremendous ways in the past few years and this technology is used in medical applications to solve so many problems, some of the innovations, and the problems are discussed and the importance of the technology is explained here. VLSI technology will give more and we discovered that the growth and trends in various industries are intertwined as a result of this report. In the field of healthcare, technological advancements have exploded. Because of the shrinking of devices and the growth in the number of biomarkers, there is a necessity for efficient processing and data storage, which VLSI technology can provide. On the other hand, the massive volumes of data created in health care must be safely stored and processed. The desire for memory devices that are lower in size and have a higher density necessitates transistor scaling.

Chapter 8: The vehicular networks face numerous security concerns that still need to be addressed before their deployment. Cryptography techniques do provide the necessary robustness against these security attacks but they put a huge overhead in the system making it difficult to manage. The simpler alternative to cryptography-based techniques is by employing trust-based models. These models prevent against most of the security attacks in VANETs and do not require any dedicated infrastructure. However, trust models presented in the literature suffer from significant inconsistencies. The major issue that has been considered in this work is the initial trust bootstrapping. The majority of the trust models assign a neutral (0.5) initial trust value to a newly encountered node which is incorrect in most of the cases. So, the proposed work emphasizes on resolving this issue by deciding adaptively for different contexts by keeping in view both the objective and subjective nature of the trust evaluation. Multiple reputation values are considered for different contexts and the final decision is made based on the Brown Gibson multi-criteria decision-making model. Another contribution of this work includes the prevention of coarse grain revocation of the reputation values due to rigidity of existing trust models. For this, a new parameter, truster disposition, has been introduced which helps in including the realistic scenarios while making a trust-based decision.

Chapter 9: In summary, safety from attackers is the main attention for the VANET users. VANETs face several safety-related challenges of passengers' comfort and safety warnings for the drivers. The attackers interrupt the communication and try to get the information of communication system. They are doing so to reduce the performance of VANET communication system. The main reason behind more attacks in VANET is because VANETs do wireless communication, so it is easy for attackers to attack a VANET network. The detailed explanation of VANET attacks and their classifications like non-repudiation, authentication, availability, confidentiality, and data integrity are discussed in this chapter. Also, the possible challenges and security services are highlighted.

Chapter 10: In this chapter, the relevance of machine learning for biomedical circuits is explained, and details on different biomedical circuits using different IC technologies are provided. The chapter also provides an insight to the logic design of a novel compressor-based approximate multiplier. The approximate multiplier when used on circuits implementing neural networks for biomedical applications can provide large energy gains and increased speed. Efficient NN-based SoCs, ASIC, and FPGA-based NN accelerators and wearable NN-based devices, all require low power multiplication units as a basic unit of computation through the NN layers. An important requirement is low power, as some applications need to stand long hours of operation and at the same time for circuits that can be driven by battery for the wearable applications. Evidently, the proposed approximate multipliers can provide high-energy efficiency with lower degradation in the quality of result. A range of truncation-based designs are proposed that can be used for several customized applications.

Chapter 11: After performing the tasks, we reach a conclusion that a cross-domain valence shifter predicts better results. But there is still need to work on cross-domain algorithms to increase the accuracy of the results. For further studies, we can consider the effect of emotions on analysis of data. Emoticons have been used by some of the authors in previous studies and theyproved them to be useful in analysis with a higher rate of efficiency. In this paper, this is performed on emojis-based sentiment analysis and classifiers to gain insights into user's emotional reactions posts on social websites with the utilization of the Google CloudPrediction API to estimate the sentiment of the posts from posted texts and emoji-based reactions. Due to the lack of large-scale data sets, the prevailing approach in visual sentimentanalysis is to leverage models trained for object classification in large data sets like ImageNet. So, there is need to develop a technique for visual sentimental analysis. In future work, we tried to develop the technique for the visual sentimental analysis.

Editors

Dr. Balwinder Raj is working as Associate Professor in ECE Department, NITTTR Chandigarh since 2019. Earlier, he worked as Assistant Professor at NIT Jalandhar and IIITM Gwalior. He has more than 15 years of teaching/research experience. He did B. Tech, Electronics Engineering (PTU Jalandhar) in 2004, M. Tech-Microelectronics (PU Chandigarh) in 2006 and Ph.D-VLSI Design (IIT Roorkee), India in 2010. For further research work, European Commission awarded him Mobility of life research fellowship for postdoc research work at University of Rome, Tor Vergata, Italy in 2010-2011. He also worked as visiting researcher at KTH University Sweden, Oct-Nov. 2013 and Aalto University Finland June 2017. Dr. Raj received best teacher's award from ISTE New Delhi, in July 2013. Dr. Raj has published more than 100 research papers in international/national journals and conferences. His areas of interest in research are Nanoscale Semiconductor Device Modeling, sensors design, FinFET based Memory design, Low Power VLSI Design.

Prof Brij B. Gupta received the PhD degree from Indian Institute of Technology (IIT) Roorkee. In more than 16 years of his professional experience, he published over 400 papers in journals/conferences including 30 books and 08 Patents with over 14000 citations. He has received numerous national and international awards including Canadian Commonwealth Scholarship (2009), Faculty Research Fellowship Award (2017), MeitY, GoI, IEEE GCCE outstanding and WIE paper awards and Best Faculty Award (2018 & 2019), NIT Kurukshetra, etc. Prof Gupta is also serving as Distinguished Research Scientist with LoginRadius Inc., USA which is one of leading cybersecurity companies, especially in the field of customer identity and access management (CIAM). He is also selected in the 2021 and 2020 Stanford University's ranking of the world's top 2% scientists. He is also a visiting/adjunct professor with several universities worldwide. He is also an IEEE Senior Member (2017) and also selected as 2021Distinguished Lecturer in IEEE CTSoc. Dr Gupta is also serving as Member-in-Large, Board of Governors, IEEE Consumer Technology Society(2022-204). Prof. Gupta is also leading IJSWIS, IJSSCI and IJCAC, IGI Global, as Editor-in- Chief. Moreover, he is also serving as lead-editor of a Book Series with CRC, World Scientific and IET press. He also served as TPC members and organized/special session chairs in ICCE-2021, GCCE 2014-2021 and TPC Chair in 2018 INFOCOM: CCSNA Workshop and Publicity Co-chair in 2020 ICCCN. Dr Gupta is also serving/served as Associate/Guest Editor of IEEE TII, IEEE TITS, IoT, IEEE Big Data, ASOC, FGCS, etc. At present, Prof. Gupta is working as Director, International Center for AI and Cyber Security

Research and Innovations, and Full Professor with the Department of Computer Science and Information Engineering (CSIE), Asia University, Taiwan. His research interests include information security, Cyber physical systems, cloud computing, blockchain technologies, intrusion detection, AI, social media and networking.

Dr. Jeetendra Singh is currently working as an assistant professor in the Department of Electronics and Communication Engineering at National Institute of Technology, Sikkim. Also, he has worked in the same position in the department of ECE at NIT Kurukshetra, Haryana, from 2012 to 2013 and at NIT Jalandhar, Punjab from 2013 to 2015. He received his B.Tech degree in Electronics and Communication Engineering from Uttar Pradesh Technical University, Lucknow, India, in 2009 and the M.Tech degree in Microwave Electronics from the University of Delhi, New Delhi, India, in 2012. He received his Ph.D. degree in VLSI Design from Dr. B. R. Ambedkar NIT Jalandhar. Dr. Singh has published more than 23 research papers in reputed international journals and also he has published several national/international conference papers and book chapters in the recognized societies. His research interest includes memristive devices, novel advanced semiconductor devices, TFETs, Analog/Digital VLSI Design, MOS Device-based sensors design, etc.

1

Design and Analysis of Charge Plasma-Based SiGe Vertical TFET for Biosensing Applications

Shailendra Singh and Sanjeev Kumar Bhalla
Department of Electronics and Communication Engineering, PSIT Kanpur

Jeetendra Singh
Department of Electronics and Communication Engineering, NIT Sikkim

Shilpi Gupta
Department of Electronics and Communication Engineering, IET Lucknow

Balwinder Raj
Department of Electronics and Communication Engineering, NITTTR Chandigarh

N.K. Yadav

Fraunhofer Institute for Photonic Microsystems, Dresden, Germany

Contents

1.1	Introduction	2
1.2	Architecture and Device Simulation Setup	5
1.3	Results and Discussion	7
	1.3.1 Drain Current Analysis on Dielectric Constant and Charged Biomolecules	8
	1.3.1.1 Drain Current Is Affected By Cavity Length Variation	12
	1.3.1.2 Drain Current Is Affected By Cavity Thickness Variation	12
	1.3.2 Changes in Drain Current Sensitivity (DCS) as a Function of Various Parameters	13
	1.3.2.1 Effect of Cavity Length on DCS	15
	1.3.2.2 Effect of Cavity Thickness on DCS	15
	1.3.3 Conclusion	16
References		16

1.1 INTRODUCTION

The constant downsizing of semiconductor devices will result in a variety of nanoscale short channel phenomena. Various devices have been presented in the era of nanoscale devices to discover an alternative in leading to a reduced subthreshold slope of less than 60 mV/decade [1–3]. In leading to a reduction of subthreshold slope and low OFF current, the TFET is proven to be an outstanding device [4,5]. The total current ratio of Ion/Ioff will grow as a result of the low OFF current. Although, the TFET technology has a low current limit, this can be overcome by employing techniques such as dual gate material, heterojunction architecture at the tunnelling junction, work function optimization, and others [6–8]. These strategies are also beneficial in reducing ambipolar conduction, which is a disadvantage of the design [9–11]. Further concern with TFET production is the expensive price of ion implantation as well as the large thermal budget required for the high temperature thermal annealing method [12,13]. Because TFET is based on the Band to Band Tunneling (B2BT) concept and uses the PIN arrangement as an abrupt junction, it can be used in a wide range of applications [14].

The PIN architecture is a strongly doped device that struggles from challenges such as random dopant fluctuation (RDFs), which raises the cost of production while also increasing leakage current in the circuit because of such a broad change in the subthreshold voltage [15]. Diffusion occurs for the source and drain regions in an effort to allow the abrupt junctions, which is a tough task in and of itself. All of the aforementioned flaws are addressed by employing doping-free procedures [16–18]. Charge plasma techniques and electrostatic doped techniques are the two types of doping-free techniques. In order to construct various device junctions, both of these strategies employed separate work functions. In terms of following the appropriate working of the device using the charge plasma approach, one must assume that the silicon substrate will be kept inside the Debye length $\left(\frac{(\varepsilon_{si} V_T)}{q} + 60.N\right)^{1/2}$ [19]. The silicon substrate's carrier concentration, thermal voltage, dielectric constant, and silicon electronic charge are all represented in this equation by N. Most of those strategies are aimed at improving the device's ion current practicality and inhibiting ambipolar conduction [20,21]. Dual gate, for example, is a well-integrated doping-free device for forming distinct junctions and minimizing ambipolarity in reverse gate biasing. Two separate gate electrodes are used in the dual material; first the auxiliary gate work function M1 and another the main gate work function M2. The auxiliary gate is greater than the tunnelling gate in the tunnelling gate's work function [22–24]. Pollution detection and biological hazard detection for early system identification and treatment has recently become a serious problem. Throughout this study, we look at the suggested technology for detecting label free electrical biomolecules like DNA and protein cells using a biosensor. The Befgveld company created the first biosensors, which were developed on ion-sensitive FETs (ISFET) [25]. ISFET, on the other hand, was considered to be more important to charge biomolecules than to neutral biomolecules [26]. This system also works in conjunction with the CMOS (complementary metal oxide semiconductor) device. To address these issues, the Dual Metal FET was invented; nevertheless, this technology has the downside of sizing, which causes the effective gate capacitance to vary, resulting in significant energy usage due to parameter changes, as well as a long detection time with poor sensitivity. Dual Metal TFETs centered on charge plasma have improved their sensitivity, reaction time, and proficiency over FET-based biosensors, making them a capable technology for detecting label-free chemicals. The device's ambipolarity was suppressed, and the ON current was increased, thanks to dual metal technology.

A biosensor with such a long life span that can be used to detect disease and investigate hazardous chemicals in the environment. The nano FET is an extremely effective research tool due to its mass manufacture and endurance. This effect occurs when nanogaps are filled with neutral or charged biomolecules, causing the oxide capacitance to alter, leading to changes in drain current and threshold voltage. Many proteins, such as biotin streptavidin (k = 2.1), gluten, zein, keratin (k = 5–10), and charged amino acids (Glu, Lys, Arg, and Asp) (k = 11–26), have dielectric constants that are similar [27,28]. By wiping off a little portion of the gate dielectric material and refilling the nanogap, this biosensing can be accomplished. As a result, when biomolecules are trapped in nanogaps, the dielectric constant of the filling gaps varies, causing the drain current of the devices to alter as well.

The dielectric-modulated FET's primary functioning mechanism is based on the effective coupling between the gate and channel. The relative importance of biomolecules causes variation in electrical characteristics, which are subsequently utilized to assess the sensitivity of neutral and charge biomolecules such as bio-streptavidin and

DNA (deoxyribonucleic acid) [29]. This can be accomplished by simulating a duplicate situation with an insulation value that is comparable to the dielectric constant of biomolecules. With several short channel effects, nevertheless the FET is limited by the subthreshold value of 60 mV/decade. As the base voltage cannot be raised much further, the device's performance suffers. Several publications and research analyses have been carried out in trying to find a remedy and offset these flaws. By introducing a B2BT method with a lower subthreshold slope, tunnel FET is discovered to be a viable contender for developing alternative FETs and overcoming their disadvantages.

The TFET, on the other hand, has its own disadvantages, such as low ON current with ambipolar conduction in reverse bias states. Further approaches, such as a dual metal gate with a modified work function, gate overlap and underlap conditions, and a new material heterojunction architecture at the tunnelling junction between the source and channel interface, were employed to improve ON current and decrease ambipolarity. With referring to these reported works, in addition to taking into account these advantages into one device, a junction less and charge plasma, or the combination of both the devices can be taken into consideration for the further improvements in the device performance.

The short channel effect will happen as a result of device scalability, resulting in system performance loss. When the device's channel length falls below 20 nm, the manufacturing process becomes more difficult [30]. As a conclusion, a junction less transistor (JLT) made of heavily doped silicon nanowires has been developed, which does not involve the formation of any junctions. While the electrical characteristics of a junction less field effect transistor (JLFET) are better, random dopant fluctuation (RDF) creates significant irregularity, which is a critical problem when scaled to nanoscale dimensions. As a conclusion, no doping charge plasma-based FETs were produced.

Because of its exceptional achievements in the fields of power usage, subthreshold slope (SS), and frequency responsiveness, TFETs have arisen as a new alternative for various design MOSFETs in the modern generation. This is due to the Tunnel FET's B2BT mechanism, which electrostatically controls charge carrier transit in order to boost device efficiency exponentially. Many applications were designed in order to achieve overall low energy usage by modifying some settings in the device's maintenance and modification [31]. One of the most recently discovered applications of biosensors is for label-free detection.

TFET-based biosensors have been utilized to regulate using the advantages of both dielectric and non-dielectric materials, leading to the invention of DM TFET biosensors. Tunnel FETs based on charge plasma are created by using a source/drain electrode and selecting the appropriate metalworking function to build a p + source. In the intrinsic silicon body, the total definition is n+ drain area ($n_i = 10^{16}$ cubic meters). Because there are no abrupt connections between the source/channel and drain regions, there is no random dopant fluctuation, reducing overall cost. As a result, the maker does not need to use a high-temperature diffusion and ion implantation technique. For all intents and purposes, part of reducing the manufacturing process of charge plasma over the conventional TFET will eventually come to a close.

To improve the device ON current, we used a dual material metal gate work function in this drought. In this case, the tunnelling gate is kept larger than the auxiliary gate. The voltage at the junction of M1 and M2 is abrupt due to the discrepancy in gate work functions, resulting in a rise in the drain current and transconductance of the device, as well as a drop in the drain induced barrier lowering. When dual material gate structure is adopted, short channel effects are mitigated while no other device characteristics are sacrificed. Nevertheless, using dielectric modulated field effect transistors (DM FETs) in capacitance fluctuation sensors has a few disadvantages. Charge plasma-based DM VTFET for gate underlap regions in biosensors, on the other hand, eliminates the aforementioned difficulties.

The construction of a charged plasma-based dual material gate dielectric. In a modified VTFET, there is a gate underlap (cavity). It's made to keep biomolecules immobilized. This book chapter focuses on the commercial operation as a low cost and cost-effective production at the customer's end to detect biomolecules from a medical standpoint.

1.2 ARCHITECTURE AND DEVICE SIMULATION SETUP

Figure 1.1 depicts the equivalent diagram of a SiGe heterojunction vertical TFET structure based on charge plasma in contrast to the charge plasma gate under area for label-free biomolecule detection. With non-metallurgical connections, use the intrinsic carrier concentration parameter with a value of $n_i = 1.0 \times 10^{16}$ cm^{-3}. The gadget has a 10 nm thickness and a 40 nm channel length. This device uses the gate stacking

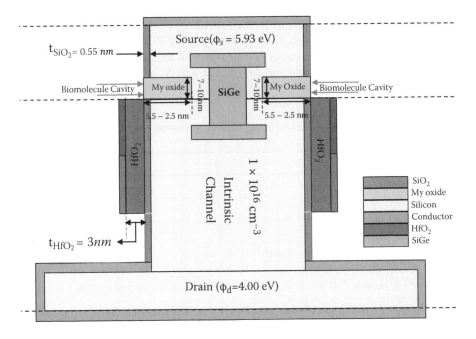

FIGURE 1.1 2-D schematic representation of dual gate charge plasma-based SiGe vertical TFET.

approach, with a high-K (HfO₂) material specification of 3 nm and a (SiO₂) material specification of 0.55 nm. It is done to prevent the issue of lattice non matching in the device body, which is silicon with SiO₂ as a gate oxide material. The 4.5 eV intake work function is used to operate the device. In addition, each source-channel tunnelling junction has a SiGe pocket with a squared width of 4 nm.

The virtual source and virtual drain regions of the charge plasma-based device are formed by adding appropriate work functions to the gate electrodes VTFET. The "p+" is made using a platinum metal electrode with a work function of 5.93 eV "In the intrinsic silicon semiconductor, there is a type source. Similarly, the "n+" is created using the hafnium work function of 3.80 eV. The device's drain area is represented by this type. One thing to remember in all of the above drafting of the proposed device is that the device thickness must be less than the Debye length, as expressed by equation (1.1) [19].

$$\left(\frac{(\varepsilon_{si} V_T)}{q} + 60.N \right)^{1/2} \tag{1.1}$$

where V_T denotes the device's thermal voltage, ε_{si} the dielectric constant, and q and N are the electronic charge carrier and silicon carrier concentration, respectively.

The schematic diagram of a charge plasma-based SiGe vertical TFFT is similar to that of a traditional device, except that the cavity region is developed to detect biomolecule immobilization. The channel is divided into two areas, designated by the labels region 1 and region 2, respectively. The first region is the biomolecule's entering space, also known as the cavity or gate underlap region (region 1), while the gate overlap region is referred as region 2. The second region (region 2) is represented by L_g as a gate overlap region with dimensions of 43 and 40 nm. For double-metal gate work functions M1 and M2, Lcavity is the breadth of the gate underlap (cavity) area, with values of 6 nm and 10 nm, and 3.8 eV and 4.5 eV, accordingly (Table 1.1).

TABLE 1.1 Device parameter of dual gate charge plasma-based SiGe vertical TFET

PARAMETERS	VALUES SPECIFIED
Channel length	40 nm
Source length	20 nm
Drain Length	20 nm
SiGe layer length	4 nm
Intrinsic doping concentration	1×10^{15} cm^{-3}
Cavity thickness	7–10 nm
Cavity width	2.5–5.5 nm
Gate work function (ϕ_{m1})	3.8 eV
Gate work function (ϕ_{m2})	4.5 eV
HfO₂ gate oxide thickness	3 nm
SiO₂ gate oxide thickness	0.55 nm

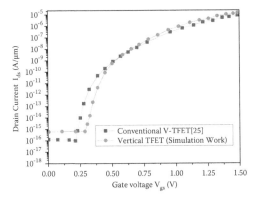

FIGURE 1.2 Calibrated graph of conventional vertical TFET [reference 25] with the simulated work at V_{ds} 1.0 V and V_{gs} 1.5 V.

Finally, Tcavity refers to the cavity width, which is calculated as 2/5 of the height-to-width ratio, which is also the gate underlap. The thickness of the spacers is set at 3 nm and 15 nm, respectively. The gate oxide thickness, tox, is gate stacked with HfO_2 and SiO_2 layers with thicknesses ranging from 0.55 to 3 nanometers. The SiO_2 layer functions as an adhesive when the Si substrate is unmasked and exposed to the atmosphere, binding the biomolecules together. The biomolecules are connected to the cavity region, which also serves as a biosensor. Silvaco TCAD simulation software is used to carry out typical simulation results for the performance evaluation of the proposed device [32]. Lombardi model and Shockley Read Hall model for recombination with Fermi-Dirac statistics with non-local B2BT were utilized to simulate the device [31]. The non-local B2BT model is also used to calculate the generation rate at the tunnelling junction. The gadget detects biomolecules by comparing the difference in dielectric constant ($K > 1$) with that of air ($K >= 1$). The gate capacitance of the vertical TFET based on charge plasma technique increases when biomolecules come into contact with the cavity, leading to an increase in the driving current, making it suited for biomolecule sensing. The calibration of the dopingless vertical TFET simulation using the testified work at V_{ds} 1.0 V is shown in Figure 1.2. Using the plot digitizer application, the data was extracted and plotted.

1.3 RESULTS AND DISCUSSION

In the natural world, there are two types of biomolecules: neutral and charged. For the neutral and changing biomolecules, a different technique to assessing the drain current characteristics is required. We just need to focus on and evaluate the dielectric constant in neutral biomolecules; however, in charged biomolecules, both the dielectric constant and charge must be studied for drain current change. As an outcome, when neutral biomolecules are considered for the proposed charge plasma-based SiGe vertical TFET in the underlap region for modelling drain current characteristics, the drain current characteristics are simulated. Charged biomolecules, on the other side, will account for both the dielectric and the altered biomolecules in the underlap region.

1.3.1 Drain Current Analysis on Dielectric Constant and Charged Biomolecules

Figure 1.3(a) and 1.5(a) exhibit the influence of neutral biomolecules on the drain current under the underlap condition with a cavity length of 7 nm at $V_{ds} = 0.5$ V and $V_{gs} = 1.5$ V, respectively. Furthermore, the cavity length of 9 nm at $V_{ds} = 1.5$ V and $V_{gs} = 1.5$ V causes a divide in the (I_{ds}) drain drive current characteristics, as seen in Figure 1.4.

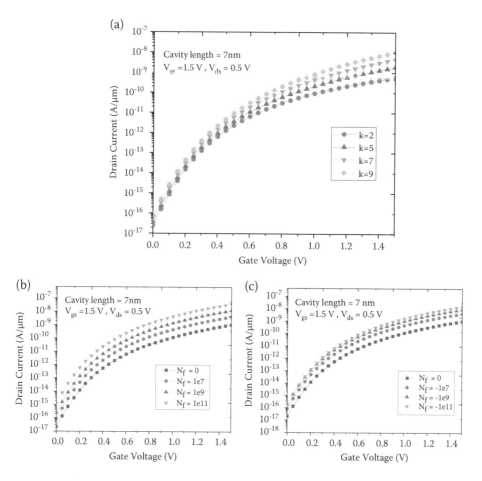

FIGURE 1.3 Drain current variation of dual gate charge plasma-based SiGe vertical TFET with cavity length thickness of 7 nm at $V_{ds} = 0.5$ V and $V_{gs} = 1.5$ V with respect to (a) dielectric constant k = (2 to 9); (b) positive charge biomolecules ($N_f = 0$ to $1e^{11}$); (c) negative charge biomolecules ($N_f = 0$ to $-1e^{11}$).

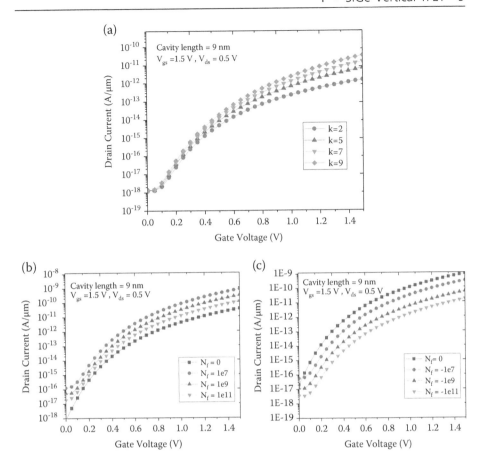

FIGURE 1.4 Drain current variation of dual gate charge plasma-based SiGe vertical TFET with cavity length thickness of 9 nm at $V_{ds} = 0.5$ V and $V_{gs} = 1.5$ V with respect to (a) dielectric constant k = (2 to 9); (b) positive charge biomolecules (Nf = 0 to $1e^{11}$); (c) negative charge biomolecules (Nf = 0 to $-1e^{11}$).

The barrier width between the channel's conduction band and the source's valence band widens as cavities are formed, resulting in a very low possibility of electron tunnelling. As a result, current conduction in this situation is low. As shown in Figures 1.3(a), 1.4(a), 1.5(a), and 1.6(a), the device ON current increases as the dielectric constant underlap condition increases.

When the dielectric constant rises, however, the OFF-current remains almost constant, resulting in significant band bending and width reduction. Figures 1.3(a) and (b), 1.4(a) and (b), 1.5(a) and (b), and 1.6(a) and (b) demonstrate the electrical properties of the drain current for the scenario where charged biomolecules are immobilized in the cavity region, respectively.

The I_{ON} current increases when positively charged proteins are present, while the I_{OFF} current drops when negatively charged biomolecules are present. Because

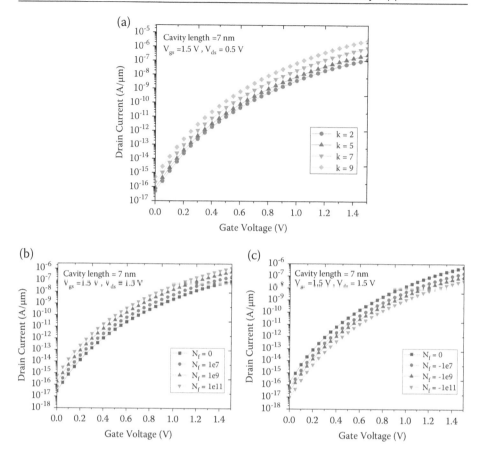

FIGURE 1.5 Drain current variation of dual gate charge plasma-based SiGe vertical TFET with cavity length thickness of 7 nm at V_{ds} = 1.5 V and V_{gs} = 1.5 V with respect to (a) dielectric constant k = (2 to 9); (b) positive charge biomolecules (Nf = 0 to $1e^{11}$); (c) negative charge biomolecules (Nf = 0 to $-1e^{11}$).

differing polarity charges are immobilized in the cavity region, the barrier width between the channel's conduction band and the source's valance band is reduced.

When the cavity is filled with biomolecules and the dielectric constant is more than 50, the model is proven to be unsuccessful. Table 1.2 shows the dielectric constants of a few biomolecules. These proteins' conductivity must first range from semiconductor to insulator.

A gate that underlaps dielectric modulated VTFET has been constructed in this study to incorporate the dielectric modulation (DM) technology for label-free electronic detection based on biomolecules such as aminopropyl triethoxysilane, biotin, protein, uricase, DNA, enzyme, and others as listed in Table 1.2 with their various values [33–35]. We investigated a gate underlap arrangement with the trap condition of the biomolecules for the charge plasma-based SiGe VTFET. In order to account for the dielectric modulation technique for detection of label free biomolecules such as

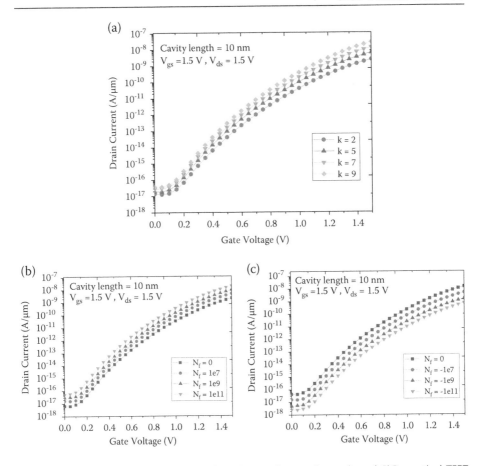

FIGURE 1.6 Drain current variation of dual gate charge plasma-based SiGe vertical TFET with cavity length thickness of 9 nm at $V_{ds} = 1.5$ V and $V_{gs} = 1.5$ V with respect to (a) dielectric constant k = (2 to 9); (b) positive charge biomolecules (Nf = 0 to $1e^{11}$); (c) negative charge biomolecules (Nf = 0 to $-1e^{11}$).

TABLE 1.2 Dielectric constant equivalence for biomolecules

NAME OF BIOMOLECULE	EQUIVALENT K VALUE
Aminopropyl triethoxy-silane	3.570
Biotin	2.630
Protein	2.500
Uricase	1.540

aminopropyl triethoxysilane, biotin, protein, uricase, DNA, enzyme, and so on, this book chapter focuses on charge plasma-based SiGe vertical TFET. We will investigate the underlap distribution of my oxide for this proposed device, in which the oxide region is uncovered due to interactions between biomolecules and the provided cavity.

As seen in Figures 1.3 and 1.4, the drain current varies dramatically from Nf = 10^{11}cm^2 to -10^{11}cm^2. The ON-current increases when the cavity is filled with positively charged biomolecules; however, when the cavity is filled with neutral biomolecules, the OFF-current remains almost constant (dielectric constant K = 7). The device based on underlap charge plasma condition of SiGe vertical TFET exhibits contradicting behaviour for drain current when negative charge density grows, resulting in drain current deterioration, as seen in Figures 1.2c and 1.3c. Figures 1.3, 1.4, 1.5, and 1.6 indicate that the drain current achieves its maximum value when the cavity length is 7 nm, and it reaches its minimum value when the cavity length is 9 nm.

1.3.1.1 Drain Current Is Affected By Cavity Length Variation

Figures 1.7(a) and 1.7(b) depict the drain current variation as a function of cavity length. As the cavity length rises from 7 to 9 nm, the ON current falls while the gate width remains constant at 40 nm. The following features demonstrate that the fluctuation in drain current at V_{ds} = 1.5 V is minimal when compared to the drain current at V_{ds} = 0.5 V. The slight change in drain current exhibited as a response to varying cavity length is attributed to the fact that charge plasma oriented SiGe vertical TFETs reflect the tunnelling principle, whereas ordinary DM JLTFETs do not. FET is based on the diffusion concept.

1.3.1.2 Drain Current Is Affected By Cavity Thickness Variation

Figures 1.8(a) and 1.8(b) show the drain current versus gate bias for different cavity thickness (tcavity) values while maintaining the channel length constant at 40 nm. When V_{ds} = 1.5 V is compared to V_{ds} = 0.5 V, the drain current changes little. The deviation in drain current with respect to V_{gs} (gate source) voltage reduces as tcavity

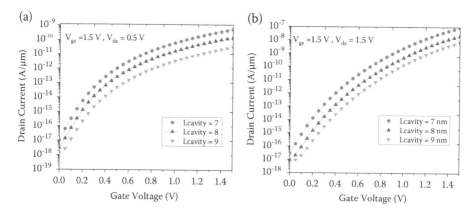

FIGURE 1.7 Drain current variation of dual gate charge plasma-based SiGe vertical TFET for cavity length thickness form 7 to 9 nm at V_{gs} = 1.5 V: (a) V_{ds} = 0.5 V and (b) V_{ds} = 1.5 V.

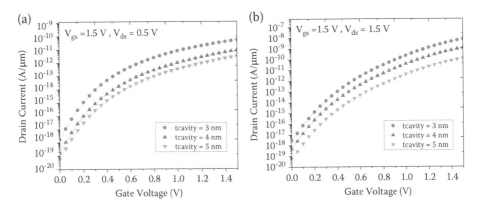

FIGURE 1.8 Drain current variation of dual gate charge plasma-based SiGe vertical TFET for cavity length width form 3 to 5 nm at V_{gs} = 1.5 V: (a) V_{ds} = 0.5 V and (b) V_{ds} = 1.5 V.

(cavity thickness) increases from 3 to 5 nm. Figure 1.8 shows that the drain current achieves its maximum value when the cavity thickness tcavity is 3 nm. Figure 1.8 shows the impact of adjusting the tcavity on the I_{ds} versus V_{gs} characteristics. Higher tunnelling barriers at the source channel interface due to the introduction of a SiGe pocket layer, which lowers the tunnelling barrier for high V_{gs} values, due to the reduction in drain current as the tcavity rises.

1.3.2 Changes in Drain Current Sensitivity (DCS) as a Function of Various Parameters

Biosensors' DCS may be used to evaluate their performance: The DCS factor ID and I_{bioD}, given by equation (1.2), respectively, represent the drain current levels while the cavity is empty and filled with a dielectric substance. The charge density of biomolecules is influenced by the various values of neutral biomolecules and the dielectric constant.

$$DCS(\%) = \left(\frac{(I_d - I_{Biod})}{I_d} * 100 \right) \quad (1.2)$$

Figure 1.9 shows the DCS in terms of gate voltage fluctuation for two underlap cavity lengths of 7 and 9 nm, respectively.

Figure 1.10 shows DCS parameters for neutral biomolecules with varying charge densities for V_{ds} = 1.5 V at 7 nm cavity length and 9 nm cavity length for various dielectric constants. When the dielectric constant of the cavity regions is gradually increased from air to enhanced levels, i.e., K >1, biomolecules are trapped in the cavity. The DCS is proportional to the value of the dielectric constant, as seen in the

14 Advanced Circuits and Systems for Healthcare and Security Applications

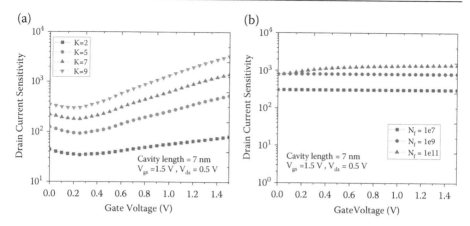

FIGURE 1.9 DCS variation of dual gate charge plasma-based SiGe vertical TFET with cavity length thickness of 7 nm at $V_{ds} = 0.5$ V and $V_{gs} = 1.5$ V with respect to (a) dielectric constant k = (2 to 9) (b) positive charge biomolecules ($N_f = 1e^7$ to $1e^{11}$).

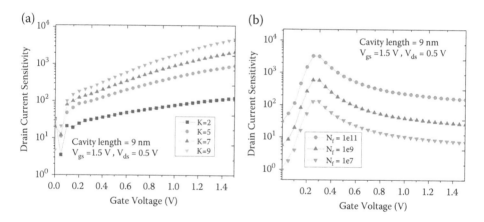

FIGURE 1.10 DCS variation of dual gate charge plasma-based SiGe vertical TFET with cavity length thickness of 9 nm at $V_{ds} = 0.5$ V and $V_{gs} = 1.5$ V with respect to (a) dielectric constant k = (2 to 9) (b) positive charge biomolecules (Nf = $1e^7$ to $1e^{11}$).

diagram. DCS rises with increasing positive biomolecule concentrations, as seen in Figures 1.9(b) and 1.10(b).

It also causes a shift in the drain current and potential, demonstrating the influence of DCS of SiGe heterojunction vertical TFETs based on charge plasma as a proportional behavior to anticipate the existence of biomolecules in the cavity. The factor of DCS rises as the the channel width increases, as shown in Table 1.3. Long channel length devices, on the other hand, have weak sensing capabilities. It is evident that a low gate bias leads to a significant improvement in DCS for various dielectric constant values and charge density levels.

TABLE 1.3 DCS as a function of dielectric constant and positive biomolecules

DIELECTRIC CONSTANT	DCS (DRAIN CURRENT SENSITIVITY)			
	$V_{DS} = 0.5V$		$V_{DS} = 1.5\ V$	
	CAVITY (7 NM)	CAVITY (9 NM)	CAVITY (7 NM)	CAVITY (9 NM)
K = 2	79.35475	118.8055	110.9494	172.3501
K = 5	528.2201	887.066	307.2841	599.322
K = 7	1418.786	2131.238	1274.889	1148.17
K = 9	3273.416	4672.41	4891.461	2242.795

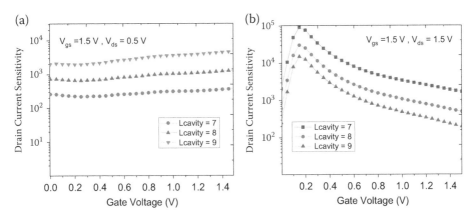

FIGURE 1.11 DCS variation of dual gate charge plasma-based SiGe vertical TFET for cavity length thickness form 7 to 9 nm at $V_{gs} = 1.5$ V (a) $V_{ds} = 0.5$ V and (b) $V_{ds} = 1.5$ V.

1.3.2.1 Effect of Cavity Length on DCS

Figure 1.11 shows how DCS changes with the drain-source voltage $V_{ds} = 0.5$ and 1.5 V. When Lcavity varies, however, the reader can notice a tiny variation in DCS. As seen in Figure 1.11, the DCS variation will grow as the gate source voltage increases. The sensitivity reaches its greatest value at a cavity length of 9 nm. The sensitivity reaches its greatest value at a cavity length of 9 nm. The DCS of charge plasma-based pocket SiGe vertical TFETs is not greatly influenced by cavity length since the tunnelling mechanism is utilized, but in the case of traditional FETs, the diffusion process is used for channel transmission.

1.3.2.2 Effect of Cavity Thickness on DCS

Figure 1.12 shows the DCS variation as a function of cavity thickness tcavity. The DCS voltage with respect to the gate voltage falls as the tcavity (cavity thickness) increases from 3 to 5 nm. The DCS displays the greatest value when the cavity thickness is 3 nm, but the minimum value when the cavity thickness is 5 nm in DCS.

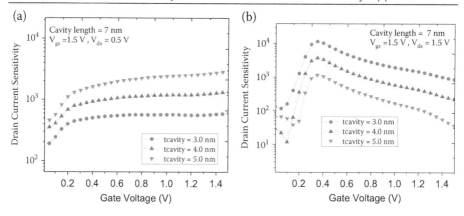

FIGURE 1.12 DCS variation of dual gate charge plasma-based SiGe vertical TFET for cavity length width form 3 to 5 nm at V_{gs} = 1.5 V (a) V_{ds} = 0.5 V and (b) V_{ds} = 1.5 V.

Because the tunnelling barrier at the source-channel boundary rises with cavity length (tcavity), the DCS decreases, resulting in reduced sensing of the proposed device charge plasma-based SiGe heterojunction vertical TFET.

1.3.3 Conclusion

We have analyzed charged and neutral biomolecule detection with the suggested charge plasma-based SiGe heterojunction vertical TFET for the underlap region in this book chapter. This draught stands out to the readers by distinguishing itself from other sensing devices currently on the market by demonstrating its greatest sensitivity in detecting charge and neutral biomolecules. As a result, the proposed device demonstrates improved performance for cost-effective biomedical diagnosis instrument construction, as it achieves increased sensitivity through a moderate optimum selection of geometrical parameters such as cavity length and thickness at the tunnelling junction interface with appropriate biassing conditions. The biasing voltage V_{gs} is held constant at 1.5 V, whereas V_{ds} varies between 0.5 and 1.5 V. After analyzing its variation to different cavity dimension parameters with dielectric constant to its electrical characteristics, it can be concluded that the proposed charge plasma-based SiGe vertical TFET has a low leakage device current and is a highly sensitive factor for the change and neutral biomolecules.

REFERENCES

[1] Tanaka, J., T. Toyabe, S. Ihara, S. Kimura, H. Noda, and K. Itoh. "Simulation of sub-0.1-mu m MOSFETs with completely suppressed short-channel effect." *IEEE Electron Device Letters* 14, no. 8 (1993): 396–399.

[2] Bricout, P. -H., and E. Dubois. "Short-channel effect immunity and current capability of sub-0.1-micron MOSFET's using a recessed channel." *IEEE Transactions on Electron Devices* 43, no. 8 (1996): 1251–1255.

[3] Singh, S., and B. Raj. "Vertical tunnel-fet analysis for excessive low power digital applications." In 2018 First International Conference on Secure Cyber Computing and Communication (ICSCCC), pp. 192–197. IEEE, 2018.

[4] Shockley, W. "The path to the conception of the junction transistor." *IEEE Transactions on Electron Devices* 31, no. 11 (1984): 1523–1546.

[5] Singh, S., and B. Raj. "Study of parametric variations on hetero-junction vertical t-shape TFET for suppressing ambipolar conduction." *Journal of Pure & Applied Physics* 58 (June 2020): 478–485.

[6] Singh, S., and B. Raj. "Analytical and compact modeling analysis of a SiGe heteromaterial vertical L-shaped TFET." *Silicon* 14, no. 5 (2021): 2135–2145.

[7] Moore, G. E. Cramming more components onto integrated circuits, Electronics 38, no. 8 (1965): 1–6, April 19.

[8] Frank, D. J., R. H. Dennard, E. Nowak, P. M. Solomon, Y. Taur, and H.-S. P. Wong. "Device scaling limits of Si MOSFETs and their application dependencies." *Proceedings of the IEEE* 89, no. 3 (2001): 259–288.

[9] Roy, K., S. Mukhopadhyay, and H. Mahmoodi-Meimand. "Leakage current mechanisms and leakage reduction techniques in deep-submicrometer CMOS circuits." *Proceedings of the IEEE* 91, no. 2 (2003): 305–327.

[10] Singh, S., and B. Raj. "Modeling and simulation analysis of SiGe heterojunction double gate vertical t-shaped tunnel FET." *Superlattices and Microstructures* 142 (2020): 106496.

[11] Edgar, L. J. "Method and apparatus for controlling electric currents." U.S. Patent 1,745, 175, issued January 28, 1930.

[12] Singh, S., and B. Raj. "Two-dimensional analytical modeling of the surface potential and drain current of a double-gate vertical t-shaped tunnel field-effect transistor." *Journal of Computational Electronics* 19, no. 3 (2020): 1154–1163.

[13] Rabaey, J. M., A. P. Chandrakasan, and B. Nikolic. "The devices." Digital integrated circuits: a design perspective, 2nd edn. Pearson Education, New York (2003).

[14] Singh, S., and B. Raj. "Design and analysis of a heterojunction vertical t-shaped tunnel field effect transistor." *Journal of Electronic Materials* 48, no. 10 (2019): 6253–6260.

[15] Dennard, R. H., F. H. Gaensslen, L. Kuhn, and H. N. Yu. "Design of micron MOS switching devices, IEDM." *Digest of Technical Papers* 15 (1972): 344.

[16] Singh, S., M. Khosla, G. Wadhwa, and B. Raj. "Design and analysis of double-gate junctionless vertical TFET for gas sensing applications." *Applied Physics A* 127, no. 1 (2021): 1–7.

[17] Singh, S., and B. Raj. "Analytical modelling and simulation of Si-Ge hetero-junction dual material gate vertical T-shaped tunnel FET." *Silicon* 13 (2021): 1139–1150.

[18] Wadhwa, G., P. Kamboj, and B. Raj. "Design optimisation of junctionless TFET biosensor for high sensitivity." *Advances in Natural Sciences: Nanoscience and Nanotechnology* 10, no. 4 (2019): 045001.

[19] Bala, S., and M. Khosla. "Design and analysis of electrostatic doped tunnel CNTFET for various process parameters variation." *Superlattices and Microstructures* 124 (2018): 160–167.

[20] Gupta, S., S. Wairya, and S. Singh. "Analytical modeling and simulation of a triple metal vertical TFET with hetero-junction gate stack." *Superlattices and Microstructures* 157 (2021): 106992–10705.

[21] Kavalieros, J., B. Doyle, S. Datta, G. Dewey, M. Doczy, B. Jin, D. Lionberger et al. "Tri-gate transistor architecture with high-k gate dielectrics, metal gates and strain

engineering." In 2006 Symposium on VLSI Technology, 2006. Digest of Technical Papers, (2006): 50–51. IEEE.
[22] Singh, S., and B. Raj. "Analysis of ONOFIC technique using SiGe heterojunction double gate vertical TFET for low power applications." *Silicon* 13, no. 7 (2020): 2115–2124.
[23] Peesa, R. B., and D. K. Panda. "Rapid Detection of Biomolecules in a Junction Less Tunnel Field-Effect Transistor (JL-TFET) Biosensor." *Silicon* 14, no. 4 (2021): 1705–1711.
[24] Singh, S., and B. Raj. "Design and Analysis of I ON and Ambipolar Current for Vertical TFET." In *Manufacturing Engineering*, 541–559. Springer, Singapore, 2020.
[25] Nigam, K., P. Kondekar, and D. Sharma. "High frequency performance of dual metal gate vertical tunnel field effect transistor based on work function engineering." *Micro & Nano Letters* 11, no. 6 (2016): 319–322.
[26] Singh, S., and B. Raj. "Analytical modeling and simulation analysis of T-shaped III-V heterojunction vertical T-FET." *Superlattices and Microstructures* 147 (2020): 106717.
[27] Aslam, M., D. Sharma, S. Yadav, D. Soni, N. Sharma, and A. Gedam. "A comparative investigation of low work-function metal implantation in the oxide region for improving electrostatic characteristics of charge plasma TFET." *Micro & Nano Letters* 14, no. 2 (2019): 123–128.
[28] Devi, W., Vandana, B. B., and P. D. Pukhrambam. "N+ pocket-doped vertical TFET for enhanced sensitivity in biosensing applications: modeling and simulation." *IEEE Transactions on Electron Devices* 67, no. 5 (2020): 2133–2139.
[29] Wangkheirakpam, V., Devi, B. B., and P. D. Pukhrambam. "N+ pocket doped vertical TFET based dielectric-modulated biosensor considering non-ideal hybridization issue: A simulation study." *IEEE Transactions on Nanotechnology* 19 (2020): 156–162.
[30] Singh, S., S. Yadav, and S. K. Bhalla. "An Improved Analytical Modeling and Simulation of Gate Stacked Linearly Graded Work Function Vertical TFET." *Silicon* 13, no. 7 (2021): 1–14.
[31] Wang, P.-F., K. Hilsenbeck, T. h. Nirschl, M. Oswald, C. h. Stepper, M. Weis, D. Schmitt-Landsiedel, and W. Hansch. "Complementary tunneling transistor for low power application." *Solid-State Electronics* 48, no. 12 (2004): 2281–2286.
[32] Manual, ATLAS User'S. "Device simulation software." Silvaco Int., Santa Clara, CA (2008).
[33] Paliwal, A., M. Tomar, and V. Gupta. "Complex dielectric constant of various biomolecules as a function of wavelength using surface plasmon resonance." *Journal of Applied Physics* 116, no. 2 (2014): 023109.
[34] Densmore, A., D. -X. Xu, S. Janz, P. Waldron, T. Mischki, G. Lopinski, A. Delâge et al. "Spiral-path high-sensitivity silicon photonic wire molecular sensor with temperature-independent response." *Optics letter* 33, no. 6 (2008): 596–598.
[35] Verma, S. K., S. Singh, G. Wadhwa, and B. Raj. "Detection of Biomolecules Using Charge-Plasma Based Gate Underlap Dielectric Modulated Dopingless TFET." *Transactions on Electrical and Electronic Materials* 21, no. 5 (2020): 528–535.

Borospherene-Based Molecular Junctions for Sensing Applications

2

Jupinder Kaur, Ravinder Kumar, and Ravinder Singh Sawhney

Department of Electronics Technology, Guru Nanak Dev University, Amritsar, Punjab, India

Contents

2.1	Introduction	20
2.2	Applications of B_{40} as Sensors	22
2.3	Transport Phenomenon in Boron Fullerene Molecular Junctions	24
2.4	Methodology	25
2.5	Results for Transport Phenomenon in Boron Fullerene Molecular Junctions	27
	2.5.1 Current-Voltage Curve	27
	2.5.2 Transmission Spectra	29
	2.5.3 Transmission Pathways	29
	2.5.4 Transmission Eigen States	32
	2.5.5 Differential Conductance	34
2.6	Conclusion	35
Acknowledgments		35
References		35

DOI: 10.1201/9781003189633-2

2.1 INTRODUCTION

The field of molecular electronics is highly favorable for commencing experiments to design memory and logic devices, which are very small in contrast to the devices fabricated utilizing the famous silicon technology [1]. These devices can be effectively utilized in various applications such as switches, dielectrics, transistors, rectifiers, and memories. Carbon-based fullerenes are an excellent choice to design molecular devices for incorporating these applications. C_{60} fullerene became a significant material in designing molecular junctions due to its superconductivity and NDR behavior [2]. Soon smaller carbon fullerenes began to be investigated and the synthesis of C_{20} in 2000 brought a revolution in the field of carbon fullerenes [3]. This led to significant progress in the field of molecular junctions. An et al. scrutinized Li decorated C_{20} [4] and Otani et al. reported that the conduction in a C_{20} junction is dependent on the number of fullerenes attached to the electrodes [5]. Density functional theory was used to explore the transport behavior in second period elements doped C_{20} [6]. Baei et al. investigated the electronic properties of transition elements doped C_{20} fullerene [7]. Although extensive research is still in progress in the field of carbon fullerenes, yet boron-based fullerenes have garnered the interest of the research community due to their exceptional properties. With the synthesis of all-boron fullerene B_{40} in the laboratory by Zhai et al. in 2014, this field gained great impetus [8]. It is a highly stable and robust molecule with D_{2d} symmetry with structure analogous to the Chinese red lantern (Figure 2.1). It has intertwined two convex caps along with four ribbons of B_{10} and eight B_9 ribbons positioned vertically and horizontally, respectively. The highest occupied molecular orbital (HOMO) and lowest unoccupied molecular orbital (LUMO) gap for this fullerene is 3.13 eV, which is quite close to the famous Buckminster fullerene having a HOMO-LUMO gap of 3.02 eV. It has a smaller diameter in contrast to the carbon-based fullerene as a result of which Y, Ca, and La prefer to align at endohedral-position of this fullerene. It is suitable for capturing and storing hydrogen atoms (up to 16 atoms).

    R. He and his team revealed that utilizing density functional theory calculations is better than C_{60} as it showcases more Raman active modes [9]. It was deduced that it

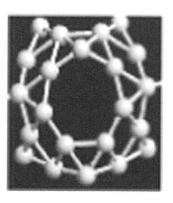

FIGURE 2.1   Borospherene.

comprises of 43 Infrared and 73 Raman active modes. Zhang et al. investigated the opto-electronic properties of this fullerene and inferred that it is an excellent choice for utilization in electronic applications as it demonstrates negative differential resistance under non-equilibrium conditions [10]. The encapsulation of metal atoms like Na, Li, Tl, K, and Ba do not affect the stability of this fullerene [11]. Barium and sodium have greater stability in the cage. On the other hand, potassium, thallium, and lithium choose to align at exohedral position above the heptagonal rings. The encapsulation of alkali metal and super-alkali molecule comprising of oxides of Li and K alter the optical properties of B_{40} [12,13] while that of transition metal elements like Fe, Co, Mg, and Ni modifies the spin-electronic behavior of B_{40} [14]. Mg and Fe encapsulated B_{40} has spin polarized transmission spectra while Ni-decorated borospherene has unpolarized one and in case of Co encapsulation, it is unpolarized for heptagonal ring interaction and polarized for hexagonal ring interaction. In addition to this B_{40} is a highly aromatic molecule [15] and also exhibits super-atomicity due to the presence of 2s, 2p, 2d, 2f orbitals [16]. The transport properties of B_{40} are alterable on doping with other elements like sulphur, aluminium, phosphorous, and silicon [17]. Doping phosphorous in the boron fullerene enhances the transport phenomenon while that of silicon, sulphur, and aluminium decreases the conductivity values. Yang et al. utilized the density functional theory to probe the smaller boron fullerenes viz., B_{28}, B_{38}, and B_{40} [18]. All three small fullerenes portray aggregation behavior. In contrast to B_{28} and B_{38}, B_{40} showcased the best aggregation behavior. As a result, it can be used as an assembled material. Further, a study has been conducted by placing the complex series C_6H_6Cr outside the B_{40} cage with Cr sandwiched between the center of the hexagonal and heptagonal holes [19]. These complexes are entirely stabilized by C_6H_6Cr fragments, as independent coordination sites depict almost the same values of coordination energies. These complexes have similar coordination patterns as that of dibenzene chromium and hence, can be used in the design of unique complexes. Keyhanian et al. scrutinized the electronic structures and properties of M-doped and M-encapsulated B_{40} (M = Mn and Fe) in gas and water phases using density functional theory calculations [20] and deduced that doping and encapsulation of the transition metals is an excellent and effective method for improving the properties of the fullerene in both the phases. The encapsulation of the cobalt atom in the B_{40} cage alters its magnetic properties [21]. Co-doped B_{40} showcases magnetic movement having a binary alternative between 2.98 µB and 1.00 µB. Therefore, based on the binary magnetic moments of Co@B_{40}, it is a promising candidate for single molecular spin electronic devices. Shakerzadeh and co-workers elucidated the electronic and non-linear optical properties of M@B_{40} (where M = Sc, Ti, V, Cr, Mn, Fe, Co, Ni, Cu, Zn) [22]. The transition metal atoms were positioned outside the heptagonal and hexagonal holes. It was deduced that decorating the fullerene with transition metals reduces the HOMO-LUMO gap. Co@B_{40} displays a much greater NLO response in comparison to Li@B_{40}. Luo et al. investigated the synergy of anti-cancer drug cisplatin with lithium doped X_4B_{32} (where X = C, Si) [23]. From the adsorption energy values, it is inferred that lithium alone cannot increase the tendency of nanocluster to adsorb the cisplatin drug. It was found that utilizing carbon along with Li@X_4B_{32} increases the adsorption energy of cisplatin. Therefore, it is a promising candidate for drug delivery applications. Hamadi and co-workers used density functional theory calculations and examined iron-anchored

B_{40} fullerene (Fe@B_{40}) [24]. It was deduced that iron atom chemisorbs on both hexagonal and heptagonal rings but the adsorption energy for Fe atom was greater over the heptagonal ring in contrast to the hexagonal ring. For a CO oxidation reaction, Fe@B_{40} is suggested as a highly active noble metal. Kosar and team probed boron fullerene for its ability as an anode material for sodium-ion batteries as energy storage materials [25]. The synergy of boron cage with both Na and Na+ was examined and small cell voltages were attained, whereas these cell voltages were increased by encapsulating halides. It was found that placing Na metal atoms outside the B_{40} nanocage and halides inside the nanocage influence the electrochemical properties of X@B_{40}. Among the various metal atoms, fluoride (F) depicts the electronegativity, which results in strong binding with Na+, resulting in an increase in cell voltage.

2.2 APPLICATIONS OF B_{40} AS SENSORS

In the last few years, B_{40} has been extensively utilized in various sensing applications as it has large surface area and encompasses both acidic and basic sites on its surface as a result of which it can detect both bases and acids. Density functional theory calculations were utilized by Dong et al. to explore the hydrogen storage capacity of borospherene encapsulating transition metals. [26]. Tin exhibited extraordinary bonding with B_{40} and 6Ti@B_{40} could effectively store 34 hydrogen molecules. This process requires very little energy and hence is reversible. B_{40} can also sense carbon dioxide from a mixture of hydrogen, nitrogen, and methane [27]. Two fullerene molecules can capture 12 atoms of gas and hence it is an appropriate candidate for separating and storing carbon dioxide. Further investigations have been done to enhance the CO_2 detection capability of borospherene by encapsulation of Li atom in the cage in order to reduce the energy barrier considerably resulting in enhanced adsorption of the gas [28]. B_{40} is also an excellent molecule for ammonia adsorption [29]. The conductivity of borospherene is enhanced on adsorption of ammonia gas it lowers the HOMO-LUMO gap. The ammonia molecule breaks into its constituents NH and H_2 on the surface of borospherene. C. Tang et al. investigated the hydrogen storage inB_{40} encapsulated with Sc [30]. The Sc atom had more stability outside the holes. A single Sc atom could effectively adsorb up to 5H-molecules and hence, Sc@B_{40} is an excellent choice for storing hydrogen gas. Z. Maniei et al. explored the nitrogen dioxide detection ability of B_{40} along with Li decorated fullerene [31]. Both the considered variants were found to be capable of adsorbing nitrogen dioxide. The electronic characteristics of the fullerene were altered on adsorption of NO_2. The adsorption of the gas decreases the energy gap in pure fullerene and increases the gap in its Li variant. Hence, both the variants can be employed as sensors for detecting NO_2. B_{40} is also an excellent nanomaterial to be employed as biomarker for predicting the sequence of nucleobases in a DNA strand as the interaction of B_{40} with various DNA nucleobases gives different conduction properties [32]. The borospherene molecule has also been used as a nanocarrier for delivery of drugs like hydroxyurea by amino acid functionalization of the fullerene [33]. B_{40} is an excellent choice to design sensors to detect toxic sulphur gases

H_2S and SO_2 [34]. Figure 2.2 portrays the adsorption of both gases on the heptagonal and hexagonal rings of the borospherene. The adsorption phenomenon was considered for both the rings because the gas molecule has the tendency to get adsorbed at any location on the surface of the adsorbent. It was inferred that H_2S is physisorbed on the heptagonal ring while it is chemisorbed on the hexagonal ring of the borospherene. SO_2 on the other hand is chemisorbed on the both rings and is found to dissociate on the surface of the fullerene. The transmission spectra and density of states analysis depict that LUMO plays the most significant role in transmission in the devices formed by utilizing the structures given in Figure 2.2(a, c, d) while HOMO is dominant in case the device is formed by using the structure given in Figure 2.2(b). The I-V curve shows that the value of current is different for all the devices and hence borospherene is an excellent material to be utilized in detection of lethal sulphur gases. Also, the values of recovery time for all the structures are very small and hence borospherene can be employed as a reversible sensor for detecting these gases.

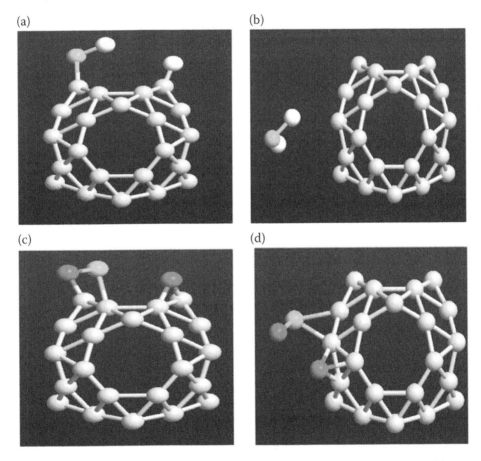

FIGURE 2.2 Optimized structures (a) H_2S chemisorbed on six-membered ring (b) H_2S Physisorbed on seven-membered ring (c) SO_2 chemisorbed on six-membered ring and (d) SO_2 chemisorbed on seven-membered ring of B_{40} [34].

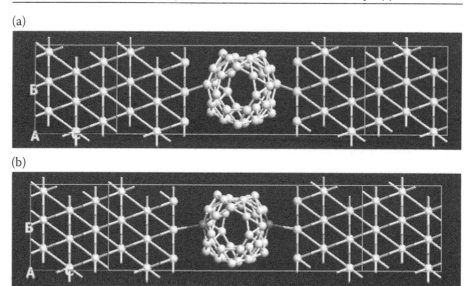

FIGURE 2.3 (a) Pure borospherene device; (b) borospherene device with radon/radium at the anchor positions (dark atoms represent radioactive elements) [35].

B_{40} is also a suitable material to detect the existence of radioactive elements [35]. To explore the utilization of borospherene in detecting the existence of radium and radon, a two-probe device is formed by replacing the anchor atoms with both radioactive elements one by one, as shown in Figure 2.3. This is because the impact of doping is most pronounced at this position. The transport calculations shows that the current is significantly decreased when borospherene is doped with radium/radon. Therefore, it is a suitable material to sense radioactive elements.

2.3 TRANSPORT PHENOMENON IN BORON FULLERENE MOLECULAR JUNCTIONS

From previously published works it is deduced that for molecular junctions, the transport phenomenon is reliant on humongous factors, such as the material utilized for formulating the electrodes, intrinsic features of the central molecule, length of the molecular junction, the alligator clips, stress, temperature, orientation, and delocalization of the HOMO and LUMO along with the energy gap present between them [36,37]. A lot of research has already been conducted in this field to garner knowledge about the impact of these variations on the transport behavior of molecular junctions [38,39]. Oshima et al. investigated the effect of variegated electrode orientations on the transport properties and inferred that due to influence of azimuthal deviations, gold exhibits asymmetrical orientations [40]. The effect of variation of electrode orientations was further scrutinized and published [41]. Although much research has already been done in exploring the various factors affecting transmission in case of single molecule junctions yet the comparison of a molecular junction formed by suspending a molecule in between two electrodes and attaching the molecule to the leads thus forming a bridge, still remains an area yet to be explored. So, here, we have explored and compared the transport phenomenon in borospherene-based molecular junction by forming two devices; firstly, by simply suspending the molecule in between two leads (without bridge) and then secondly by attaching the fullerene molecule to the leads using single bond on both sides (with bridge).

2.4 METHODOLOGY

Density functional theory (DFT) [42] is utilized to execute the imperative calculations in the regime of non-equilibrium Green's function (NEGF) [43] using Atomistix Tool Kit (ATK) software [44]. Generalized gradient approximation (GGA) suggested by Perdew, Burke, and Ernzerhof (PBE) is utilized as the exchange-correlation. Since the type of basis set employed for the simulations plays a decisive role in determining the transport behavior of the molecular junction under study, we employ double zeta-polarized (DZP) basis set as it has been reliably implemented to inquire molecular junctions with different electrode orientations [45].

We devise two molecular junctions utilizing borospherene and gold electrodes with <1, 1, 1> orientation (Figure 2.4). In the first molecular junction, the B_{40} molecule is simply suspended between the two electrodes (without bridge) while in the second case the B_{40} molecule is attached with the gold electrodes via a single atom of boron on both sides (with bridge). The junctions thus formed constitute the left and right electrodes with the scattering region positioned between them. Gold is the optimum choice for electrodes due to its high stability at room temperature. <1, 1, 1> electrode orientation is preferred as a nanowire demonstrates superior conduction with this configuration. To enhance coupling between the B_{40} and the electrodes, an extended molecule is considered by taking certain fragments of the electrode in the scattering region on each side.

The elastic scattering probability included in the Landauer-Imry formula is employed along with NEGF to comprehend the transport phenomenon of the molecular junctions in the low bias region. Self-consistent solution of NEGF in order to formulate the electron density matrix followed by determination of the electrostatic potential is helpful in calculating the current [46]. The molecular junctions thus formulated are utilized to contemplate the various transport properties.

The current-voltage relation of a molecular junction is chracterized using the Landauer-Buttiker formula [46]. It takes advantage of the probability of transmission {T(E,V)} to calculate the current using the following equation

$$I(V) = 2\frac{e}{h}\int_{\mu_L}^{\mu_R} T(E, V)[f(E - \mu_L) - f(E - \mu_R)]dE \qquad (2.1)$$

(a)

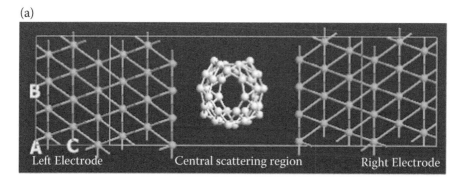

(b)

FIGURE 2.4 B_{40}-based molecular junction (a) without bridge and (b) with bridge. Here, dark grey atoms denote gold atoms while white atoms represent boron atoms.

where T (E, V) is the transmission function and μ_R/μ_L represent the potential of the right and left electrode, respectively. The determination of transmission function requires the knowledge of two parameters viz. the extent of coupling amongst the central molecule and the metallic leads and the location of the molecular energy levels. In this work, a similar method is adopted to determine the transmission, as given below:

$$T^k(E, V) = T_r[\Gamma_1^k(E, V)G_M^k(E, V)\Gamma_2^k(E, V)G_M^{k\prime}(E, V)] \qquad (2.2)$$

Here, Γ (E) denotes the coupling function. This function is capable of providing adequate information about the type of interaction between the central molecule and the electrodes. G_M (E,V) represents the Green's function.

It is presumed that the molecular energy levels have the propensity to drift up with a value eV_{mol} that is in close proximity with the molecular electrostatic potential. For the purpose of simplicity, it is assumed that the energy levels are fixed while the substrate on the other hand has a predisposition to drift downwards by eV_{mol}. Therefore, the value of potentials of both leads is contemplated in the following manner:

$$\mu_L = E_F - eV_{mol} = E_F - \eta eV \qquad (2.3)$$

$$\mu_R = E_F - eV + eV_{mol} = E_F + (1 - \eta)eV \qquad (2.4)$$

Here, e denotes the charge on electron, E_F is the equilibrium Fermi energy, and η is advantageous in illustrating the manner in which the potential difference (V) is shared amongst the two leads. Its value is numerically calculated from: $-\eta = V_{mol}/V$. From earlier executed experiments utilizing scanning tunneling microscope, it is inferred that by appropriately arranging the contacts in a well-proportioned position, η is fine-tuned to around 0.5. This parameter is an essential parameter in illustrating the profile of central molecules in molecular devices since the electrostatic field existing inside the molecule has negligible impact on the conductance curve.

Next, the differential conductance is calculated, as given below:

$$G = \frac{I(V_R, T_L, T_R) - I(V_L, T_L, T_R)}{V_R - V_L} \qquad (2.5)$$

2.5 RESULTS FOR TRANSPORT PHENOMENON IN BORON FULLERENE MOLECULAR JUNCTIONS

2.5.1 Current-Voltage Curve

The current-voltage curve being one of the most imperative transport properties of a molecular junction is calculated first at various voltages in the range of −1V to 1V with a step size of 0.5 V (Table 2.1). As per Lang and Kohn, the work function is closely related to the density of states and the electrons participating in conduction. A large value of work function implies less chances of transmission and vice versa. To settle this issue, more energy has to be provided. From Figure 2.5, we visualize that for borospherene device with bridge; the I-V curve is symmetrical for both the positive as well as negative bias and the amplitude of current increases with increase in bias. The highest value of current is 164284.76 nA at 1 V, which implies that the value of work function is very

TABLE 2.1 Values of current at variegated bias for both the molecular junctions of B_{40}

VOLTAGE (V)	B_{40} WITH BRIDGE (NA)	B_{40} WITHOUT BRIDGE (NA)
−1	−167227.37	−4361.65
−0.5	−83593.75	−2359.88
0	0	0
0.5	82778.69	2125.36
1	164284.76	6412.59

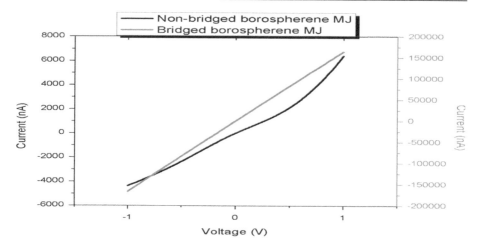

FIGURE 2.5 Comparative current-voltage curve for B_{40} molecular junctions with and without bridge formation.

small. The linearity in the current-voltage curve highlights the metallic nature of the molecular junction. The slope of the curve is visualized to decrease with increase in bias. For borospherene without bridge formation, the value of current at all the considered voltages is comparatively very low with a maximum value of 6412.59 nA at 1 V. The slope of curve changes abruptly in the considered range with the highest value perceived between 0.5 V and 1 V. The curve is non-linear from 0.2 V to 0.8 V.

To further analyze the conduction phenomenon in the considered devices, we contemplate the rectification ratio and summarize it in Table 2.2. The linearity of the I-V curve can be easily represented by the rectification ratio (RR). Very minute disparities that cannot be visualized from the I-V curve are clearly observed from the values of rectification ratio. For borospherene molecular junction with bridge formation, the rectification ratio is almost constant as the current-voltage curve portrays linearity while for the molecular junction without bridge a large value of rectification ratio is demonstrated for 1 V. The variation in RR with change in bias is because of different behaviors of the orbitals of the device under the positive and negative bias [47]. The value of rectification ratio is large when the forward current is large and the reverse current is small while a low rectification ratio is obtained with a small forward current and large reverse current. As RR is very large for a borospherene device without bridge formation, this device is an apt choice to be employed as a rectifier [48].

TABLE 2.2 Rectification ratio of both B_{40} molecular junctions

I(V)/-I(V)	BOROSPHERENE WITH BRIDGE	BOROSPHERENE WITHOUT BRIDGE
0.5/−0.5	0.99	0.90
1/−1	0.98	1.47

2.5.2 Transmission Spectra

The transmission spectra utilize the molecular orbital energies to give information about the type of transmission through the molecular junction. In case of an ideal periodic structure, the transmission spectra are calculated as a sum of all the modes available at every energy value in the spectrum. The number of peaks in the transmission spectra plays a significant role in defining the amount of transmission. More peaks are responsible for more transmission. In addition, the sharpness and width of the peaks are important factors in determining the amount of coupling between the central molecule and the metallic leads. The transmission spectra for both the considered borospherene molecularjunctions at various voltages are as shown in Figure 2.6 for each voltage. 0 eV is considered the fermi energy. At equilibrium, for a borospherene device with a bridge, a higher magnitude of transmission is indicated, implying larger transmission. Also, the peaks are broader in this device, thus indicating a larger extent of coupling between the central region and electrodes. The higher magnitude of transmission and broader peaks imply an easier flow of charger carriers from left to right metallic lead leading to more transmission. As the voltage is increased from 0 V to 1 V, the transmission peaks are seen to shift from their position. Also, when negative bias is applied, the number of peaks decreases, thus indicating lesser transmission at negative bias. At equilibrium, for a borospherene molecular junction without bridging, a smaller magnitude of transmission peaks is obtained in comparison to a borospherene molecular junction with a bridge. Also, the peaks obtained are sharper and have less width. The magnitude, height, and width of the peaks imply that the amount of transmission is comparatively less in this molecular junction. With an increase in applied voltage, the peaks shift from their position and their magnitude decreases. When negative voltage is applied, the transmission peaks shift away from the fermi level. This implies a decrease in transmission. Thus, from the analysis of transmission spectra, it is concluded that bridged-borospherene device supports more transmission in comparison to borospherene molecular junction without bridge formation. These results are in accordance with the current-voltage curve as described previously.

2.5.3 Transmission Pathways

The transmission pathways provide useful information about all the possible paths available for transmission across the borospherene molecular junction when a potential difference is applied between the two metallic leads. Figure 2.7(a-b) showcases the transmission pathways when both the left and right electrodes are at 0 V. The transmission pathways in this case are a result of asymmetrical chemical potential between the two leads. The various transmission pathways are shown using grey colored arrows. An arrow is drawn only when the local transmission magnitude amongst two atoms is at least 40% of the highest value of local transmission at that particular energy. The arrow is represented with a blue color in the case of positive transmission between two atoms when the second atom is present along the direction in which transmission occurs. The red arrows represent the component of conduction, which lies in the reverse direction and is responsible for reduction in net

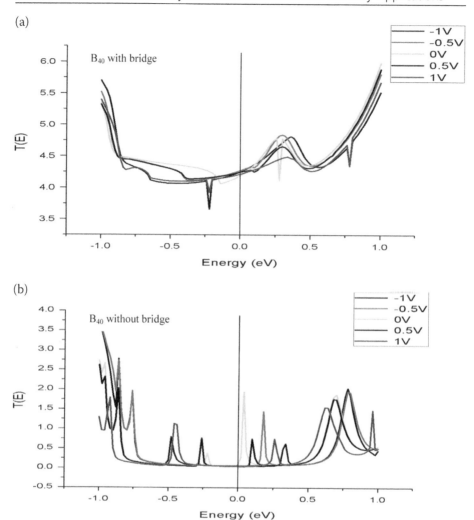

FIGURE 2.6 Transmission spectra at various voltages for B_{40} (a) with bridge and (b) without bridge. 0 eV represents the fermi level.

current. The purple arrows, usually visualized in a perpendicular direction, assay the carrier scattering phenomenon in the molecular junction.

From Figure 2.7, it is clearly visualized that for a bridged-borospherene molecular junction, more transmission pathways are present in comparison to the other device. In addition, more blue arrows and fewer red arrows are depicted by borospherene MJ with a bridge, thus implying more transmission in the forward direction leading to more conduction. In contrast, for a molecular junction with suspended borospherene, a larger number of red arrows and very few blue arrows mean that the reverse component of current significantly decreases the total transmission through the device.

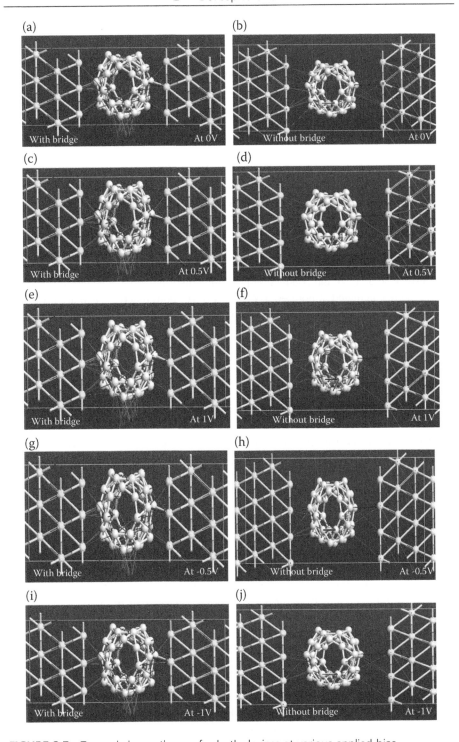

FIGURE 2.7 Transmission pathways for both devices at various applied bias.

To get better insight into the conduction phenomenon of the aforesaid molecular junction, the transmission pathways are calculated for both negative and positive biases. Figure 2.7(c-d) illustrates the transmission pathways for both the molecular junction when the left electrode is at a chemical potential of 0.5 V. In the case of a bridged junction, the number of forward pathways is seen to increase; thus, a stronger current is depicted by the I-V curve (Figure 2.5). For the suspended borosphrene device, the number of transmission pathways in the forward direction is seen to decrease drastically. This is indeed the reason for the non-linearity in the I-V curve (Figure 2.5).

Figure 2.7(e-f) depicts the transmission pathways for both molecular junctions at 1 V. The forward transmission pathways are found to increase further in the case of a molecular junction with bridged borospherene, thus supporting the fact that at 1 V the device assays the highest value of current. For the other device, there is significant increase in both the forward and reverse transmission pathways. This is the reason the I-V curve is almost linear around 1 V and the device portrays maximum current at this voltage.

Further, to analyze the conduction phenomenon at negative bias, the transmission pathways are calculated at both −0.5 V and −1 V for both junctions and are illustrated in Figure 2.7(g-j). From the figure, it can be inferred that the molecular functions behave in a similar way as in the case of a positive bias. Thus, the analysis of transmission pathways gives a clear picture about the conduction and supports the results deduced from current-voltage curve obtained previously.

2.5.4 Transmission Eigen States

Next, to identify the transport phenomenon through borospherene-based molecular devices, we further analyze and compare the Eigen states of both devices considered. The effect of oxidation on the transport channels of a molecular function can easily be observed from the transmission Eigen states because the transmission is defined as the addition of Eigen states of all the transport channels, which participate in transmission. At 0 V (Figure 2.8(a-b)), the transmission Eigen states of bridged borospherene molecular junction are more delocalized with both the sigma and pie bonds playing a dominant role in transmission while for a suspended borospherene molecular junction, the delocalization is less and more oriented on the left side of the scattering region with sigma bonds dominating the little transmission that is occurring through the junction. As delocalization of molecular orbitals is more enhanced in the case of a bridged molecular junction, it assays more current as already mentioned earlier.

To analyze the role of molecular orbitals in transmission to a greater extent, the transmission Eigen states are determined for both the positive and negative bias, as shown in Figure 2.8(c-j). At 0.5 V, for a bridged device, more delocalization is visible with more dominance of sigma bonds, whereas for borospherene MJ without a bridge, the delocalization is limited to the left side with pie orbitals coming into the picture alongside sigma orbitals. At 1 V, bridged-borospherene MJ demonstrates more delocalization with the complete dominance of sigma bonds while the other device shows very little delocalization with pie bonds clearly dominating the transmission.

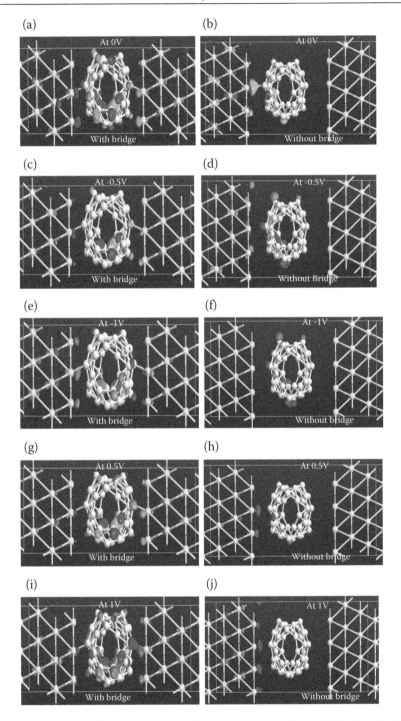

FIGURE 2.8 Transmission Eigen states for B_{40} molecular junctions with/without bridge at various voltages.

For a negative bias of −0.5 V, a bridged device assays more delocalization with sigma bonds responsible for nearly all the transmission through the device, whereas non-bridged MJ gives very little delocalization mostly due to sigma orbitals, thus implying that at −0.5 V the transmission is less in the case of both devices. For −1 V, bridged-borospherene MJ portrays delocalization of its molecular orbitals to a large extent with participation of both the sigma and pie orbitals while non-bridged MJ assays have very little delocalization on the left side due to sigma orbitals.

Thus, on analyzing the transmission Eigen states at all biases, we infer that the reason for more current in borospherene MJ with a bridge is the more delocalization of the orbitals in this device leading to more transmission due to enhanced coupling between the central molecule and the leads on both sides. Thus, these results are quite in agreement with the I-V curve, transmission spectra, and the transmission pathways calculated previously.

2.5.5 Differential Conductance

Next, the differential conductance is determined for both considered devices (Figure 2.9). It is clearly portrayed in the figure that bridged-borospherene device showcases larger values of differential conductance in comparison to the non-bridged device. In addition to this, bridged-borospherene MJ depicts the Kondo Tunneling phenomenon with a value of differential conductance approaching $2G_0$ ($G_0 = 2e^2/h = 77.48$ μS) as a consequence of equal dissemination of the central molecule with both the left and right electrodes. At equilibrium, the value of conductance is found to be 2.15 G_0, thus indicating lifting of Coulomb Blockade, paving ways for utilizing this device in SET (single electron transistor) applications. The results are in close co-relation with all the results obtained previously.

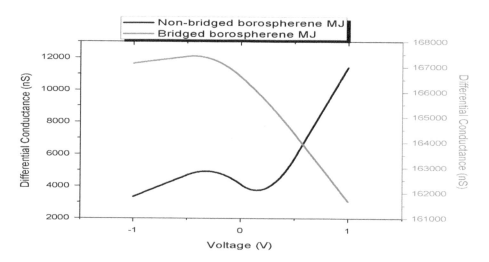

FIGURE 2.9 Comparative graph of differential conductance for both molecular junctions.

2.6 CONCLUSION

Two borospherene-based molecular junctions are designed, firstly by simply suspending the fullerene between the two electrodes (no bridge) and then by attaching the fullerene via single bonds with both electrodes (with bridge). The current-voltage curve, transmission spectra, transmission Eigen states, transmission pathways, and differential conductance are determined for both molecular junctions, taking into consideration the density functional theory and non-equilibrium Green's function formula. Borospherene molecular junction with a bridge demonstrates superior transport behavior with the highest value of current, 164284.76 nA at 1 V. Non-bridged borospherene MJ on the other hand showcases a large rectification ratio (1.47), making it an ideal choice to be exploited in rectifier applications. The superior transport phenomenon in the case of bridged-borospherene MJ is due to more coupling between the fullerene and electrodes, a greater number of forward transmission pathways and enhanced delocalization of molecular orbitals with participation of both the pie and sigma bonds in transmission. Also, this device assays Kondo Tunneling phenomenon with differential conductance approaching $2G_0$. At equilibrium, the conductance is $2.15G_0$, thus remarking the lifting of coulomb blockade and paving ways for exploring this single molecule device in SET (single electron transistor) applications.

FUNDING: This research did not receive any specific grant from funding agencies in the public, commercial, or not-for-profit-sectors.

ACKNOWLEDGMENTS

The authors are grateful to Virtual Nano Lab at Guru Nanak Dev University, Amritsar, India for providing computational facilities.
Declarations of Interest: None

REFERENCES

[1] C Joachim, JK Gimzewski, A Aviram, *Nature*, 408, 541 (2000).
[2] P Zhao, DS Liu, HY Liu, SJ Li, G Chen, *Organic Electronics*, 14(4), 1109 (2013).
[3] H Prinzbach, A Weiler, P Landenberger, F Wahl, J Worth, LT Scott, MGelmont, D Olevano, BV Issendorff, *Nature*, 407, 60 (2000).
[4] YP An, CL Yang, MS Wang, XG Ma, DH Wang, *Current Applied Physics*, 10(1), 260 (2009).
[5] M Otani, T Ono, K Hirose, *Physical Review B*, 69(12), 121408(R) (2004).

[6] G Singh, M Kaur, S Mahajan, RS Sawhney, *Materials Today: Proceedings*, 3(6), 2422 (2016).
[7] MT Baei, A Soltani, P Torabi, F Hosseini, *Monatsheftefür Chemie-Chemical Monthly*, 145(9), 1401 (2014).
[8] H J Zhai, Y F Zhao, W L Li, Q Chen, H Bai, H S Hu, Z A Piazza, W J Tian, H G Lu, Y B Wu, Y W Mu, G F Wei, Z P Liu, J Li, S D Li, L S Wang, *Nature Chemistry*, 6, 727 (2014).
[9] R He, X C Zeng, *Chemical Communications*, 51, 3185 (2015).
[10] Z Yang, Y L Ji, G Lan, L C Xu, X Liu, B Xu, *Solid State Communications*, 217, 38 (2015).
[11] B Lin, H Dong, C Du, T Hou, H Lin, Y Li, *Nanotechnology*, 27, 075501 (2016).
[12] W Fa, S Chen, S Pande, X C Zeng, *The Journal of Physical Chemistry A.*, 119(45), 11208 (2015).
[13] E Shakerzadeh, Z Biglari, E Tahmasebi, *Chemical Physics Letters*, 654, 76 (2016).
[14] Z Li, G Yu, X Zhang, X Huang, W Chen, *Physica E: Low-dimensional Systems and Nanostructures*, 94, 204 (2017).
[15] W Wang, Y D Guo, X H Yan, *RSC Advances* (2016). doi: 10.1039/C6RA00179C.
[16] E V Shah, D R Roy, *Physica E*, 84, 354 (2016).
[17] J Kaur, R Kaur, 8th ICCCNT, 1–4 (2017). doi: 10.1109/ICCCNT.2017.8203969.
[18] T Yang, Y Li, D Hao, L Li, H Peng, P Jin, *International Journal of Quantum Chemistry*, 25921 (2019).
[19] L Pei, D Li, H Li, Y Mu, H Lu, Y Wu, S Li, *Jounal of Cluster Science* (2019).
[20] M Keyhanian, D Farmanzadeh, *Journal of Molecular Liquids*, 249, 111638 (2019).
[21] TZ Wen, AZ Xie, JL Li, *Chemical Physics*, 20019 (2019). doi: 10.1016/j.chemphys.2 019.110587
[22] E Shakerzadeh, M Yousefizadeh, M Bamdad, *Inorganic Chemistry Communications* (2019). doi: 10.1016/j.inoche.2019.107692
[23] Q Luo, W Gu, Novel borospherenes as cisplatin anticancer drug delivery systems. *Molecular Physics* (2020). doi: 10.1080/00268976.2020.1774088
[24] H Hamadi, E Shakerzadeh, MD Esrafili, Fe-decorated all-boron B40 fullerene serving as a potential promosing active catalyst for CO oxidation: A DFT mechanistic approach. *Polyhedron* (2020). doi: 10.1016/j.poly.2020.114699
[25] N Kosar, F Ullah, K Ayub, U Rashid, M Imran, MN Ahmed, T Mahmood, *Materials Science in Semiconductor Processing* 121 (2020) 105437.
[26] H Dong, T Hou, ST Lee, Y Li, *Scientific Reports*, 5, 9952 (2015).
[27] H Dong, B Lin, K Gilmore, T Hou, ST Lee, Y Li, *Current Applied Physics*, 15(9), 1084 (2015).
[28] G Gao, F Ma, Y Jiao, Q Sun, Y Jiao, E Waclawik, A Du, *Computational Materials Science*, 108(A), 38–41 (2015).
[29] M Moradi, V Vahabi, A Bodaghi, *Journal of Molecular Liquids*, 223, 315–320 (2016).
[30] C Tang, X Zhang, *International Journal of Hydrogen Energy*, 41, 16992–1699 (2016).
[31] Z Maniei, E Shakerzadeh, Z Mahdavifar, *Chemical Physics Letters*, 691, 360–365 (2018).
[32] J Kaur, R Kumar, Borosphere-based biomarker for DNA sequencing: a DFT study. *JournalCompututational Electronics* (2021). doi: 10.1007/s10825-021-01731-6
[33] Y Zheng, K Shan, Y Zhang et al., Amino-acid functionalized borospherenes as drug delivery systems, *Biophysical Chemistry* (2020). 10.1016/j.bpc.2020.106407
[34] J Kaur, R Kumar, R Vohra, RS Sawhney, *Journal of Molecular Modeling*, 26, 17 (2020).
[35] J Kaur, R Kumar, R Vohra, RS Sawhney, *Journal of Materials Research*, accepted July, (2020). doi: 10.1557/jmr.2020.205
[36] RA Marcus, *The Journal of Chemical Physics*, 24, 966 (1956).
[37] A Nitzan, MA Ratner, *Science*, 300(5624), 1384 (2003).
[38] H Basch, R Cohen, MA Ratner, *Nano Letters*, 5(9), 1668 (2005).
[39] Y Xue, MA Ratner, *Physical Review B*, 68, 115406-1-18 (2003).

[40] Y Oshima, K Mouri, H Hirayama, K Takayanagi, *Surface Science*, 531(3), 209 (2003).
[41] R Kaur, RS Sawhney, D Engles, *Molecular Physics*, 114(15), 2289 (2016).
[42] ND Lang, *Physical Review B: Condensed Matter and Materials Physics*, 52, 5335 (1995).
[43] J Taylor, M Brandbyge, K Stockbro, *Physical Review Letters*, 89(13), 138301 (2002).
[44] Atomistic Toolkit Manual, Quantumwise Inc. Atomistix toolkit version 13.8.0, Quantumwise A/S (http://quantumwise.com)
[45] QH Wu, P Zhao, DS Liu, G Chen, *Solid State Communications*, 174, 5 (2013).
[46] R Landauer, *Journal of Physics: Condensed Matter*, 1, 8099 (1989).
[47] G Ji, B Cui, Y Xu, C Fang, W Zhao, D Li, D Liu, *RSC Advances*, 4, 16537 (2014).
[48] A Lofgren, I Shorubalko, P Omling, AM Song, *Physical Review B*, 67(19) (2003).

Nanowire FETs for Healthcare Applications

3

Deep Singh and Amandeep Singh
Department of ECE, National Institute of Technology, Srinagar, J&K

Contents

3.1	Introduction	40
	3.1.1 MOS Transistor	41
	3.1.2 Scaling of MOSFET	41
	3.1.3 Issues at Nanoscale Level	41
3.2	Advance MOSFET Structure	42
	3.2.1 Double-Gate (DG) MOSFET	42
	3.2.2 Surrounding-Gate MOSFET	42
	3.2.3 FinFET	43
3.3	Nanowire for Biomedical Applications	43
3.4	Modeling of Nanoscale Devices	43
	3.4.1 Semiclassical Transport: Diffusive	44
	3.4.2 Ballistic Transport	45
	3.4.3 Quantum Transport	47
3.5	Modeling of Nanowire	47
	3.5.1 Electrostatic of Nanowire	48
	3.5.1.1 Physical Model	49
	3.5.1.2 Without External Force	49
	3.5.1.3 With External Force	49
	3.5.1.4 Effective Mass Approximation	50
	3.5.1.5 Poisson's Equation	50
	3.5.1.6 Quantum Electron Concentration	51

	3.5.2	Electron Mobility Modeling	52
		3.5.2.1 Low-Field Mobility Model	53
		3.5.2.2 Effective Mobility	54
	3.5.3	Carrier Transport Modeling	54
	3.5.4	Surface Potential Momodeling	55
	3.5.5	Charge Modeling	57
		3.5.5.1 Semiconductor Charge of an Undopped Nanowire	57
	3.5.6	Subband Energy Modeling	59
3.6		Healthcare Applications of Nanowire FETs	61
	3.6.1	Detection of Viruses	61
	3.6.2	Detection of Biomarkers	62
	3.6.3	Detection of DNA and RNA	64
	3.6.4	Discovery of Medication	64
3.7		Summary	65
References			65

3.1 INTRODUCTION

Nanowires are wonderful structures for current nanoelectronic devices. Many research groups are working on fundamental properties and applications of nanowires. Because of many research works, now it is possible to fabricate semiconductor nanowires with control on diameter. From the last few years, the scaling of MOSFET provides the facility of packaging a large number of field effect transistors (FETs) on a single chip. According to the prediction of Gordon Moore, the number of transistors on a single chip would double after every two years. The ITRS published that for the next decades, the scaling of CMOS technology will continue. In the last 40 years of development in IC technology, the scaling of transistors reached the nanometer dimensions and introduced 22 nm process technologies. Nowadays, the transistors are fabricated with gate lengths ranging from 22 nm to 45 nm. Intel, AMD, and IBM have demonstrated the p-channel FET with a gate length of 6 nm. Nanowires are one-dimensional structures, and very useful for the development of various nanoscale semiconductor devices such as FETs, photodetectors, LEDs, biological sensors, etc. Very soon the fabrication of transistors below 22 nm will meet its fundamental physical limitations. These physical limits made possible the application of new nanoscale devices such as nanotubes and semiconductor nanowires in new physics as the future novel electronics due to their unique electronic properties. The study of science shows that nanoscale technologies are wonderful inventions. The architecture and materials of nanoscale devices require much scaling and research. The main object of modeling and characterization of nanoscale devices is to predict and analyze the various characteristics, performance, and properties of the devices by using different modeling and simulation techniques.

3.1.1 MOS Transistor

Firstly, the working principle of FET was invented in 1930 (Lilienfeld, 1930). The MOSFET is widely used in ICs. Dimension of devices has been scaled down and complexity of circuits, the progress of IC industries is increasing explicitly. Small-scale effects include the issues of tunneling via the gate terminal and quantum confinement issue at interface, bandgap, and discrete atomistic effects (DTE) in the doping. Therefore, the problems of scattering and thermal occur due to very high power densities (HPD). The basic structure of VLSI circuits and microprocessors is the MOSFET. The gate terminal of MOSFET in CMOS technology is n-type and p-type polysilicon with silicide. This silicide reduces the series resistance of a gate terminal in CMOS technology. In CMOS technology, the circuit density is very high. Shallow trench isolation (STI) technology is used to reduce the circuit density. For the source and drains, a combination of deep and shallow implants has been used. The threshold voltage depends on doping concentration; it decreases with an increase in doping. This makes it possible to fabricate MOSFET with a shorter gate length.

3.1.2 Scaling of MOSFET

Since the invention of MOSFET, there is fast and exponential growth in integrated circuit technology. This development is based on downscaling of dimensions of MOSFET, which was first suggested in 1974. Generally, two methods of scaling are used, the constant voltage scaling method and the constant field scaling method. The increment of the packing density of the MOSFET device has been continued from more than 45 years. The ITRS published that, the length of the gate terminal will be scaled down to 7 nm by the year 2018 (ITRS, 2004). As the dimensions of the device reached the nanoscale, the various issues in the modeling and simulation of transport of carriers, the quantum mechanical effects has been induced. Available models of transport of carriers for modeling and simulation of devices, one mostly derived from Boltzmann transport equation (Roosbroeck, 1953; Lundstrom, 2000). For the last three years, the performance of transistors increasing with the reduction in size, supply voltage, and gate oxide thickness. Now dimensions of the MOS transistor start approaching the nanometer regime, so new effects need to be considered from the quantum physics phenomenon. The scaling and modification of device is necessary to keep up the improvement continuously. As the dimension of a transistor approaches the nanometer regime, the various effects collectively called short channel effects.

3.1.3 Issues at Nanoscale Level

When the length of the channel reduces beyond 50 nm, fabrication techniques require more inventions and development to deal with various issues due to various physical phenomenons. At a nanoscale, the classical-quantum physics approach fails to downscale the dimensions of the transistor, so new problems arise, such as short channel effects. Current tunneling issues due to thin oxide, from source to drain the quantum mechanical tunneling of charge carriers (Stadele, 2002; Wang & Lundstrom, 2002), due to quantum confinement the issue of increment in threshold voltage (Majima et al., 2000) are most often mentioned issues at the nanoscale level.

Dimensions of materials are responsible for the determination of various properties. A practically general method for the fabrication of nanowires with a diameter less than 10 mm has not been developed until now. A technology node less than 10 nm requires novel inventions in the transport of charge carriers in dimensional structures such as nanotubes, nanowires, etc. to realize the various applications of nanowires requires more understanding of the fundamental physics of one-dimensional nanostructures.

3.2 ADVANCE MOSFET STRUCTURE

Based on the continued downscaling of field effect transistors, ITRS predicts that by 2015, the industry of semiconductor devices would require the length of the channel to be around 10 nm. With the development of new semiconductor materials, new nanoscale devices were invented that require more scaling up to 10 nm length of the gate. The development of new multigate field-effect transistors including double gate MOSFET, Pi-gate FET (Park et al., 2001), omega-gate MOSFET (Yang et al., 2002), tri-gate MOSFET (Doyle et al., 2003), and surrounding gate MOSFETs requires more development in fabrication techniques.

3.2.1 Double-Gate (DG) MOSFET

A double-gate MOSFET has two gate terminals referred to as the front gate and back gate. Between the front gate and back gate terminal, the ultra-thin layer of silicon material is sandwiched. Two different structures of DG MOSFET symmetric and asymmetric have been developed. DG MOSFET can be operated in two different modes called three-terminal and four-terminal modes of operation. The analytical and compact models of DG MOSFET have been developed by many research groups (Subramaniam et al., 2013; Han et al., 2008).

3.2.2 Surrounding-Gate MOSFET

In the surrounding-gate MOSFET, the gate terminal covers all sides of the channel. This structure of gate and channel reduces the leakage current. The various models of surrounding-gate MOSFET have been developed for the circuit simulation. A doping-less technique is used to develop a new graded channel nanowire MOSFET and a charge plasma technique is used to create a graded channel. A nanowire FET was used to propose a photosensor, and it is found that the photosensor based on nanowire FET has better performance characteristics than DG MOSFET (Jain et al., 2016). A nanowire MOSFET with three metal gates has been developed for the application of photo-sensing (Sharma et al., 2018). A comparative study shows that CNT FET gives a better performance in comparison with nanowire FET (Singh et al., 2016).

3.2.3 FinFET

FinFET is a type of DG MOSFET and the best alternative to conventional MOSFET. Fabrication of FinFET is easy due to its simple structure. Based on various experiments, many researchers demonstrate that FinFET has less effect of short channel effects in comparison with conventional MOSFET. Different structures of FinFET as symmetric and asymmetric have been developed by many research groups (Bhattacharya & Jha, 2014).

3.3 NANOWIRE FOR BIOMEDICAL APPLICATIONS

The diseases can be detected in their early stage by using sensors with sufficient sensitivity. As the size of the material decreases in the order of a nanometer, the physical and chemical properties of the material are determined by the large area of the surface to volume ratio, and the effect of quantum size, so the properties of material at nanoscale are completely different from the materials at the macro scale. With scale decreases the ratio between the volume and the area of surface increases. The large ratio between the area of surface and volume means that the target will affect the more region of sensory structure.

Furthermore, the flow of electrons in a nanostructure is limited by the presence of impact of quantization. The most important primary works for the application of nanowires in the biomedical field is to detect and characterize the chemical and biological species ranging from the diagnosis of diseases to the discovery of drugs. Semiconducting nanowires enable direct access to various types of electrical signals rather than labels.

3.4 MODELING OF NANOSCALE DEVICES

The theoretical and experimental study of field effect transistors that have a length of the channel smaller than 10 nanometers are in the current research trend and many research groups are working in this area (Chang, 2003). After the experimental demonstration of a molecular switch, the devices based on molecular electronics, we are very near to real molecular electronic devices (Choi et al., 2000). The working principle of semiconductor devices is based on the control of the movement of electrons and holes into the device structure, and the movement of both electrons and holes is considered as transport of charge carriers. When the physics of semiconductors was at starting days, in physics of condensed matter, Shockley wrote electrons and holes in a semiconductor (Chang et al., 2003). With the change in time and research trends, the most required fundamental properties and concepts of semiconductor devices were clarified and makes it possible to convert into the understanding and working knowledge of engineers working in the area of semiconductor device manufacturing. The electrons and holes with effective mass in semiconductors are considered semiclassical particles. Semiconductor materials such as silicon and gallium arsenide, etc. were used to make most of electronic devices, which have required bandgap, effective mass, and other properties. The model of drift and diffusion equations, which has an easy but accurate explanation of transport of charge carriers, was provided by semiconductor device engineers for most devices. Now the shrinking of the dimensions of semiconductor devices has reached the nanoscale level. Because of the quantum confinement of charge carriers in semiconductor materials, the properties of materials are affected. The numbers of impurity dopants in semiconductor devices are countable and very sensitive to the geometrical structure of devices at the nanoscale. Where the fundamental semiconductor devices such as MOSFETs have been scaled down to nanoscale geometry, now nanoscale semiconductor devices made from carbon nanotubes (CNT), semiconductor nanowires, and organic molecules have been invented by device engineers. The engineers working in the area of designing and manufacturing semiconductor devices need to think differently about nanoscale semiconductor devices (Anantram et al., 2008).

To understand the transport of charge carriers in semiconductor devices at the nanoscale, developers should be aware of how charge carriers behave like semiclassical particles and also need to understand the behavior of charge carriers at an atomistic scale. The purpose of this section is to provide an introduction of charge carrier transport required to understand the modeling of semiconductor devices at the nanoscale and also modeling of semiconductor nanowires, which is introduced in the next section.

3.4.1 Semiclassical Transport: Diffusive

At the beginning age of semiconductor devices, the electrons were treated as semiclassical particles that move in the devices due to an applied electric field and random scattering of potentials. Figure 3.1 shows that electrons have random motion along a path in position and momentum space (Anantram et al., 2008).

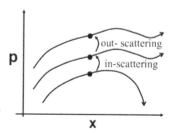

FIGURE 3.1 Random motion of electrons along a path in position momentum space (Anantram et al., 2008).

Experimentally, charge carriers in semiconductors scatter because of vibrations in the lattice, impurities due to ionized ions, and other structural defects in the lattice. The scatterings result in the charge carriers or electrons moving from one path to another, as shown in Figure 3.1. The critical dimensions of semiconductor devices should be larger than the mean free path. The mean free path is defined as the average distance between two different scattering events. Charge carriers i.e. electrons and holes have random motion into the semiconductor devices in a specific direction, which is directed by the applied electric field (Anantram et al., 2008).

To understand the scattering process, the numbers are chosen randomly when charge carriers change their path. The method is called the Monte Carlo technique, in which the average of the results of a large number of simulations is used for computational modeling for the description of the transport of charge carriers (Shockley, 1950). The researchers and device developers are interested in work with the average density of charge carriers and density of current flowing through the device, etc. For the derivation of the moment equation, a mathematical approach exists, but to the mathematical equation into an understandable form, several simplifying approaches are needed (Datta, 1996).

The phenomenological description of the transport of charge carriers can be given by the moment equation that provides deep and quantum results when all measurements are calibrated properly. If charge carriers are being accelerated at a specific location, the electron density increases with time. Electron density decreases with time because of the electron-hole recombination. The drift of electrons in the electric field causes the diffusion of a kinetic energy gradient.

Anantram et al. state that in the bulk semiconductor the mobility of electrons is energy-dependent. Momentum relaxation time is less than the energy relaxation time. The reason is small phonon energy. To thermalize and have enough carriers, some scattering events are required, and only one scattering event requires randomizing the momentum of it (Anantram et al., 2008). Figure 3.2 shows the average velocity versus electric field of electrons in silicon (Figure 3.3).

3.4.2 Ballistic Transport

The semiconductor device shown in Figure 3.4 has a ballistic region connected to two leads. The lead on the left side is labeled as the source and the lead on the right

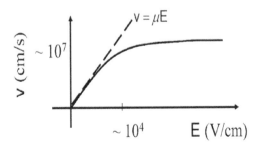

FIGURE 3.2 Average velocity and electric field of electrons in silicon (Anantram et al., 2008).

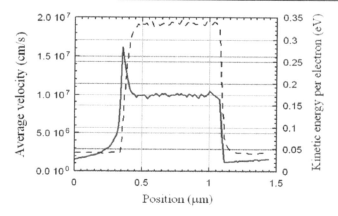

FIGURE 3.3 The average velocity and kinetic energy versus position for electrons (Anantram et al., 2008).

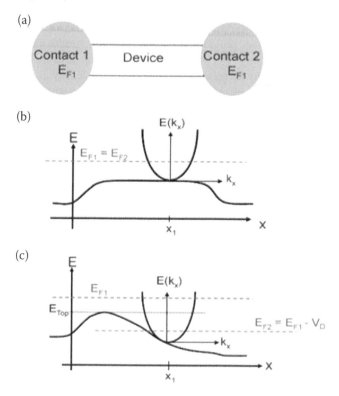

FIGURE 3.4 Structure of a semiconductor device with ballistic transport (Anantram et al., 2008).

side is labeled as the drain. The source is used to inject a flux of thermal equilibrium of charge carriers into the semiconductor devices. Both the source terminal and drain terminal are considered perfect absorbers. The two methods are used to calculate the density of electron, density of the current, and average velocity of electrons. In the first method, the charge carriers are considered as semi-classical particles, and the distribution function is obtained by solving Boltzmann's transport equation. In the second approach, charge carriers are considered as the quantum mechanical particles.

When the bias voltage is applied to the ballistic semiconductor device, the situation has been shown in Figure 3.4. Because of the injection of two thermal equilibrium fluxes into the semiconductor device, this is too far away from the condition of equilibrium. The steady-state form of Boltzmann's transport equation is the same for a semiconductor device with ballistic transport as in equilibrium. There are two Fermi levels, so it is difficult to choose only one Fermi level that gives better results.

3.4.3 Quantum Transport

Conventionally, quantum transport is studied in two-dimensional electron gases. The nanostructures of semiconductor material have different properties that make it possible to study quantum transport. The ballistic and diffusive transport mechanism discussed in the previous section can be derived from principles of quantum mechanics (Khan et al., 1987). Quantum transport of carriers in semiconductor nanostructure are studied based on the solution of quantum transport equations. Non-equilibrium Green's function approach is designed to study quantum transport of carriers including scattering, and it is based on a quantum theory of the main body.

3.5 MODELING OF NANOWIRE

In the previous section, the various approaches of charge carrier transport have been discussed. This section is about the modeling of semiconductor nanowires. The purpose of this section is to introduce the different modeling approaches of semiconductor nanowires as charge modeling, carrier transport modeling, subband modeling, and surface potential modeling. The information given in this section is required to understand the physics, working principle, and modeling of nanoelectronic devices based on semiconductor nanowires.

To study the various characteristics of high mobility junctionless nanowire transistors, a carrier scattering model has been developed and fabricated. It is investigated that the transport of carriers can be controlled by phonon scattering (Kumar, nd). To calculate the current and voltage relationship of silicon nanowire FET, the effective mass model is used (Wang et al., 2005). To measure the current-voltage characteristics of silicon nanowire, a model based on linear relationship has developed (Reza et al., 2006). The mobility of electrons in semiconductor nanowire limited by phonon was calculated by solving the Poisson Schrödinger equation and decreasing the width of nanowire ranging from 30 to 8 nm (Ramayya et al., 2007). A model based on the effective mass theory of electrons is used for the quantum simulation nanowire FET. For the study of carrier transport, NEGF and Poisson's equation have also been solved

(Shin, 2007). The quantum transport properties, an expression for current-voltage characteristics, and capacitance-voltage relationship of semiconductor nanowires have been studied by using a quasi-analytical model (Paul et al., 2007) this model is also compatible with circuit simulator as SPICE. The effect of band structure on the quantum transport properties of semiconductor nanowires was investigated (Gnani et al., 2007). It is demonstrated that nanowire FET provides better control on downscaling issue; the transport model based on subband energy has also proposed (Fiori & Iannaccone, 2007). The transport of electrons in nanowire transistors has been studied using a model based on multi-subband Boltazman's transport equation considering quasi-ballistic transport and quantum confinement of electrons. It is investigated that the BTE based model is more accurate than the model based on NGEF (Jin et al., 2008). A model to study the surface depletion and density of doping of semiconductor nanowire has been proposed by solving Poisson's equation (Andrew et al., 2012). The method of electrical transport measurements is used to prepare a single-crystal semiconductor nanowire (Cui et al., 2001). The combination of scattering matrices and solution of BTE is used to develop a quasi-ballistic transport model of semiconductor nanowire FET (Lee et al., 2015). The thermal effects such as thermal resistance can be analyzed by using a model proposed (Jain et al., 2018). To study the behavior of nanowire FET, a model was developed considering the two-dimensional quantum confinements charge carriers (Shafizade et al., 2019). An analytical model of the nanowire was proposed for the study of output characteristics of nanowire MOSFET and to implement the circuit in the simulator (Yesayan et al., 2020). Gate capacitance and transport of charge carrier are experimentally measured by using a model based on charge-based measurement technique (Zhao et al., 2009). The effect of thermal diffusion and radiation was investigated by using a model based on a transmission line matrix. A compact model based on fundamental physics was used for the estimation of subband energies, density of charge, and capacitance in semiconductor nanowire MOSFET (Ganeriwala et al., 2017).

3.5.1 Electrostatic of Nanowire

Different issues in charge carrier transport theory can be represented by using the Fredholm integral equation:

$$f = IK(F) + \phi \tag{3.1}$$

In general, the physical quantities of interest are determined by the function of the type:

$$J(f) = (g, f) = \int_g g(x)f(x)dx \tag{3.2}$$

where the function $g(x)$ and $f(x)$ belong to a branch space x and to the adjacent space respectively $f(x)$ is the solution of equation (3.1).

3.5.1.1 Physical Model

The electron began to propagate along the semiconductor nanowire, where the characterization by two variables such as position z and component of the wave vector is k_z.

A general time-dependent electric field E(t) can be applied along the nanowire:

$$i\hbar \frac{\partial}{\partial t}\psi(r, t) = \frac{\hbar^2}{2m_o}\nabla\psi(r, t) + \phi(r, t)\psi(r, t) \qquad (3.3)$$

where i = imaginary unit

$\hbar = \frac{h}{2\pi}$ is Plank's constatnt
r = position vector
t = time
ψ = electron wave function
(r, t) = electron potential energy

3.5.1.2 Without External Force

If there is no externally applied force, then

$$\phi_r(r, t) = \phi_c(r) \qquad (3.4)$$

where ϕ_c = electron potential energy due to semiconductor crystal lattice

ϕ_c is assumed to be static and periodic following the spatial periodicity of the crystal. For a static potential energy, equation (3.3) can be simplified by writing the electron wave function:

$$\psi(r, t) = \psi(r)\xi(t) \qquad (3.5)$$

The Blooch theorem imposed the periodicity of the electron wave function, which is given by

$$\psi_i(r, k) = \frac{1}{\sqrt{\Omega}}e^{ikr}\mu_i(r, k) \qquad (3.6)$$

where $\mu_i(r, k)$ = the Blooch function

Ω = volume of the semiconductor unit cell
k = electron wave function

3.5.1.3 With External Force

When the crystal lattice potential is perturbed by the lattice vibrations, impurities, or external forces, the Hamiltonian of the Schrödinger equation can be written as the perturbation of the crystal lattice Hamiltonian:

$$H_C + \phi(r) \tag{3.7}$$

where $\phi(r)$ is the electron potential energy due to external force.

The Schrodinger equation can be written as:

$$i\hbar \frac{\partial}{\partial t}\psi(r) = [E_i(-i\nabla) + \phi(r, t)]\psi(r, t) \tag{3.8}$$

where $\phi(r)$ = potential energy due to external force

E_i = energy in the ith branch

3.5.1.4 Effective Mass Approximation

In most practical cases, just the lower $E_i(k)$ states and the conduction band are occupied by electrons. Thus, we focus on the minimum of $E_i(k)$ branches. The equation can be simplified to

$$[E_i(-i\nabla) + \phi(r)]^{iko}\psi(r) = E_n e^{ikor}\psi(r) \tag{3.9}$$

where k_o is wave vector corresponding to the minimum, E_n is Eigen value of the left hand side Hamiltonian, and ψ is envelop function. In equation (3.9), the potential due to the crystal lattice, ϕ_c, is explicitly present. It is possible to define an electron effective mass as

$$m_{ij}^{-1} = \left. \frac{\partial^2 E}{\partial k_i \partial k_j} \right|_{k_o} \tag{3.10}$$

The dispersion relation can be simplified as

$$E_i(k) = \left[E_i(k_o) + \frac{\hbar^2 k_x^2}{2m_x} + \frac{\hbar^2 k_y^2}{2m_y} + \frac{\hbar^2 k_z^2}{2m_z} \right] \tag{3.11}$$

Using the parabolic dispersion relation in equations (3.9) and (3.11) resets into the parabolic effective mass approximation Schrodinger equation:

$$\frac{\hbar^2}{2}\nabla w \nabla^T + \phi(r)\psi(r) = E'_n \psi(r) \tag{3.12}$$

3.5.1.5 Poisson's Equation

Poisson's equation relates the electrostatic potential ψ and the charge distribution in the device ρ

$$\nabla[\varepsilon(r)\nabla\psi = -\rho(r) \tag{3.13}$$

where $\varepsilon(r)$ is dielectric constant

$$\phi(r) = -q\psi(r) \tag{3.14}$$

where q is electron charge

The charge distribution, $\rho(r)$ is given by

$$\rho(r) = q[p(r) - n(r) + N_d^+ - N_a^-] \tag{3.15}$$

where p(r) is electron concentration, n(r) is concentration of holes, N_d^+ is donor ionized impurities concentration, and N_a^- is acceptor ionized impurities concentration.

3.5.1.6 Quantum Electron Concentration

To determine the quantum electron concentration, it is necessary to characterize the electron density in both real and wave vector space. In the real space, the electron distribution is fully determined by the square modulus of the electron wave function. In the wave vector space, it is necessary to count the number of available electron state and their occupancy. The wave vector density of state g(k) is just the number of k states divided by the semiconductor volume. Each k state occupies a volume in the wave vector space given by Ω_B/N where Ω_B is volume of Brillouin zone, N is number of unit cell in the real space for the given volume v, and $N = v/\Omega_C$ is unit cell volume.

Then the wave vector density of states is given by

$$g(k) = 2\frac{\Omega_B/N}{v} = \frac{(v/\Omega_C)/\Omega_B}{v} = \frac{2}{(2\pi)^3} \tag{3.16}$$

where $\Omega_B \Omega_C = (2\pi)^3$ is the factor account for the electron spin energy. The energy density of state $g(E)$ can be determined by changing the variable from k to E and using $E_i(k)$:

$$g(k_z)dK_z = g(E)dE \tag{3.17}$$

$$g(E) = gk\frac{dk}{dE} \tag{3.18}$$

The state occupancy is determined by the Fermi Dirac function:

$$f(E) = \frac{1}{1 + e^{\frac{E-E_F}{k_B T}}} \tag{3.19}$$

Being k_B is Boltzmann's constant, E_F is the Fermi level.

The electron concentration under non-parabolic dispersion relation:

$$n = \int dE \frac{1 + 2\beta_v(E - \phi_i)}{\sqrt{2m^*[E - E_i + \beta_v(E - \phi_i)^2]}} \frac{1}{1 + e^{\frac{E - E_F}{k_B T}}} \tag{3.20}$$

As per holes, and ionized impurities due to dopants, classical expression are considered:

$$p = 2\left(\frac{2\pi m_h k_B T}{\hbar^2}\right)^{3/2} e^{\frac{-E_F - E_v}{k_B T}} \tag{3.21}$$

$$N_a^- = N_a f(E_a) = \frac{N_a}{1 + \frac{1}{g_a} e^{\frac{E_a - E_f}{k_B T}}} \tag{3.22}$$

$$N_d^+ = N_d \left[1 - f(-E_d)\right] = \frac{N_d}{1 + \frac{1}{g_d} e^{\frac{E_d - E_f}{k_B T}}} \tag{3.23}$$

where m_h is effective hole mass, E_v is valance band energy, and $N_a(N_d)$, $g_a(g_d)$ and $E_a(E_d)$ are acceptor (donor) concentration level degeneracy and energy, respectively.

3.5.2 Electron Mobility Modeling

The charge carrier transport properties in semiconductor materials is studied by using the mobility of charge carriers (Granzner et al., 2014). The mobility of electrons limited by phonon is given as

$$\mu_{ph}(A) = \frac{\alpha A^\beta \mu_{max}}{\alpha A^\beta + \mu_{max}} \tag{3.24}$$

where A is the cross-section area. The mobility of electrons limited by phonon is a function of the diameter of the semiconductor nanowire (Figure 3.5 and 3.6).

The mobility of semiconductor nanowires limited by surface roughness scattering is a function of the diameter of the semiconductor nanowire.

The model of dependence of electron mobility limited by surface roughness on the cross section can be derived from the function:

3 • Nanowire FETs for Healthcare Applications 53

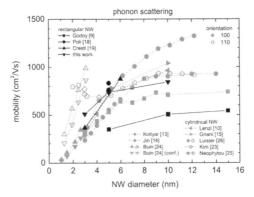

FIGURE 3.5 Mobility of electrons as a function of the diameter of the nanowire in silicon nanowires limited by phonon scattering (Granzner et al., 2014).

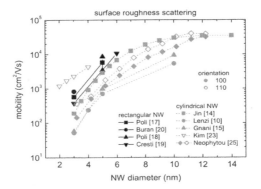

FIGURE 3.6 Mobility of electrons as a function of the diameter of the nanowire in silicon nanowires limited by surface roughness scattering (Granzner et al., 2014).

$$\mu_{sr,100}(d_{NW}) = \frac{ad_{NW}^b \mu_{max}}{ad_{NW}^b + \mu_{max}} \left(\frac{\Delta}{\Delta_{ref}}\right)^c \qquad (3.25)$$

where d_{NW} is the diameter of semiconductor nanowire (given in nanometer) and Δ is the rms roughness of the surface.

3.5.2.1 Low-Field Mobility Model

By applying Mathieson's rule, the low field mobility of charge carriers in the semiconductor nanowire can be written as

$$\mu_o^{-1} = \mu_{ph}^{-1} + \mu_{sr}^{-1} \qquad (3.26)$$

3.5.2.2 Effective Mobility

$$\mu_t(E_{eff}) = \left(\frac{E_o}{E_{eff}}\right)^p \quad (3.27)$$

where E_o and E_{eff} are fitting parameters. E_o and p are given by

$$E_o = E_1 B^{dwn}\left(1 + \frac{E_1}{E_\infty} B^{d_{NW}}\right)^{-1} \quad (3.28)$$

$$p = Cd_{NW}^\delta + p_\infty \quad (3.29)$$

The nanowire diameter is normalized by 1 nm.

3.5.3 Carrier Transport Modeling

The model of carrier transport is presented many times by using different approaches (Chia et al., 2012). The Slotboom functions express the density of electron (n) and hole (p) as

$$n = n_i e^{q\psi/kT} u \quad (3.30)$$

$$n = n_i e^{-q\psi/kT} v \quad (3.31)$$

where n_i is the intrinsic carrier concentration, ψ is the electric potential, and u and v are the Slotboom variables related to the quasi-Fermi levels of the carriers. The boundary condition of the electric potential is

$$\hat{n}\varepsilon_s \nabla \psi = q(D_d^+ - D_a^-) \quad (3.32)$$

where ε_s is the permittivity of the material, $\hat{n}$ is the unit vector normal to the surface, and D_d^+ is the surface densities (cm^{-2}) of ionized donor type and D_a^- is the surface densities (cm^{-2}) of ionized acceptor-type traps, where

$$D_d^+ = \frac{n_{dtrap} + n_i v e^{\frac{-q\psi}{kT}}}{n_i u e^{\frac{q\psi}{kT}} + n_{dtrap} + n_i v e^{\frac{-q\psi}{kT}} + p_{dtrap}} D_d \quad (3.33)$$

and

$$D_a^- = \frac{P_{atrap} + n_i u e^{\frac{q\psi}{kT}}}{n_i u e^{\frac{q\psi}{kT}} + n_{atrap} + n_i v e^{\frac{-q\psi}{kT}} + P_{atrap}} D_a \qquad (3.34)$$

The recombination rate of electrons and holes can be derive by using the Shockley-Read-Hall (SRH) recombination

$$R_{SRH} = \frac{n_i(uv - 1)}{\tau_p(ue^{q\psi/kT} + 1) + \tau_n(ve^{-q\psi/kT} + 1)} \qquad (3.35)$$

where τ_n is the electron lifetime and τ_p is the hole lifetime.

Using the Slotboom formalism, the recombination rate R_S of the charge carrier is given as

$$R_s = \frac{2Sn_i(uv - 1)}{ue^{q\psi/kT} + ve^{-q\psi/kT}} \qquad (3.36)$$

where s is velocity of surface recombination.

3.5.4 Surface Potential Momodeling

Several properties of nanowires affect surface charge density and surface potential. The surface potential of conducting material can be directly measured by using an electrochemical cell (Ganeriwala et al., 2017).

The derivation of expression for the Fermi level ($E f$) is difficult in the modeling of surface potential. The charge density of the gate is equal to the charge density of semiconductor material and can be given as

$$Q_g = q \sum_{i=1}^{N_E} \int_{E_i}^{\infty} g_{1-D}(E) F(E) dE \qquad (3.37)$$

where q represents the charge of electron, E_i represents the energy of i the subband, and N_E is the total number of subbands. g_{1-D} denotes the one-dimensional (1 − D) density of states (DOS) and is given by

$$g_{1-D} = \frac{2}{h}\sqrt{\frac{m^*}{2E}} \qquad (3.38)$$

where h denotes Planck's constant, E represents the energy, and m^* is the effective mass.

$f(E)$ in equation (3.37) denotes the Fermi–Dirac distribution function and is given as

$$f(E) = \frac{1}{1 + e^{\frac{(E-E_F)}{kT}}} \quad (3.39)$$

where k is Boltzmann's constant and T denotes the temperature. Equation (3.37) can be integrated to obtain

$$Q_g = \frac{2}{h}\sqrt{\frac{m^*kT}{2E}} \sum_{i=1}^{N_E} F_{-\frac{1}{2}} \frac{E_F - E_i}{kT} \quad (3.40)$$

$F_{-\frac{1}{2}}$ is the Fermi integral of half order. The charge density of the semiconductor and gate is also given by

$$Q_g = C_{ins}\left(V_g - V_{fb} - \frac{E_f}{q}\right) \quad (3.41)$$

For drain–source voltage, $V_{ds} = 0V$, where C_{ins} is the capacitance of insulator per unit length and V_{fb} represents the flat-band voltage. By comparing Equations (3.40) and (3.41), a single expression in terms of the Fermi level (E_F) can be given as

$$\frac{2}{h}\sqrt{\frac{m^*kT}{2E}} \sum_{i=1}^{N_E} F_{-\frac{1}{2}} \frac{E_F - E_i}{kT} = Q_g = C_{ins}\left(V_g - V_{fb} - \frac{E_f}{q}\right) \quad (3.42)$$

If the cross section of the nanowire is circular, the capacitance of insulator is given as

$$C_{ins} = \frac{2\pi\epsilon_{ins}}{\ln\left(1 + \frac{t_{ins}}{R_o}\right)} \quad (3.43)$$

where r is the radius of nanowire, ϵ_{ins} is the permittivity of the insulator, and t_{ins} is the effective thickness of the insulator.

If the cross section of a nanowire in equation (3.43) becomes,

$$C_{ins} = \frac{\epsilon_{ins} W_{ch}}{t_{ins}} \quad (3.44)$$

where W_{ch} represents the effective width of channel.

For the i the subband, a unified expression for the Fermi level can be given as

$$E_f^i = \frac{1}{2}\left(\frac{E_{f,st}}{N_E} + E_{f,at}^i - \sqrt{\left(\frac{E_{f,st}}{N_E} - E_{f,at}^i\right)^2 + \delta}\right) \quad (3.45)$$

where E_f^i denotes the Fermi level of a unified surface while considering the i in the subband only.

$\delta = 0.01$ is a smoothing parameter. The total charge density of the semiconductor and gate can be given as

$$Q_g = \sum_{i=1}^{N_E} Q_g^i = C_{ins}\left(V_{gfb} - \sum_{i=1}^{N_E} \frac{E_f^i}{q}\right) \qquad (3.46)$$

which shows that the expression of the Fermi level while considering multiple subbands can be given as

$$E_f = \frac{1}{2}\left(\frac{E_{f,st}}{N_E} + E_{f,at}^i - \sqrt{\left(\frac{E_{f,st}}{N_E} - E_{f,at}^i\right)^2 + \delta}\right) \qquad (3.47)$$

At any point in the channel for any bias, the surface potential is given by

$$\psi = \frac{E_f(V_{gc})}{q} + V_c \qquad (3.48)$$

where V_c is the channel voltage.

3.5.5 Charge Modeling

The electrostatic nanowires can be modified by controlling the charge in semiconductor layers and charge traps on the surface. The charge modeling of semiconductor nanowires has already been addressed (Ganeriwala et al., 2019). The purpose of this section is to present a generalized model of charge in semiconductor nanowires. When there is a sudden termination of semiconductor crystal, the interface state arises. The modeling of semiconductor charge in nanowires explained in this section is based on the physics of material and geometrical structure of nanowires and free from mathematical complexity (Figure 3.7).

E_f is assumed as the reference, the radius of semiconductor nanowires is denoted by R_o, and the thickness of the insulator labeled as t_{ins}. A 2D cylindrical coordinate system is used for all the calculations.

3.5.5.1 Semiconductor Charge of an Undopped Nanowire

When the semiconductor nanowire is undoped or there is no impurity added in the semiconductor nanowire, then the semiconductor charge (Q_s) can be written as a function of gate voltage (V_G) and surface potential (ϕ_s), and mathematically can be written as given:

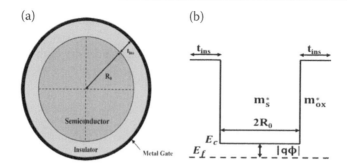

FIGURE 3.7 (a) Structure of cylindrical nanowire, (b) band diagram of the NW transistor.

$$Q_s = C_{ins}(V_G - \phi'_{ms} - \phi_s) \tag{3.49}$$

where $\phi'_{ms} = \phi_m - \chi$, ϕ_m and is work function of metal, χ is the affinity of electrons in semiconductor material, and $C_{ins} = \dfrac{2\pi \varepsilon_{ins}}{\ln\left(1 + \dfrac{t_{ins}}{R_o}\right)}$ is called the capacitance of insulator.

If Fermi energy, $E_f = 0$, then semiconductor charge density can be calculated as

$$Q_s = -q \int_0^{2\pi} \int_0^{R_o} n(r,\theta) \, dr \, d\theta \tag{3.50}$$

$$Q_s = -q \sum_{j=g_v}^{n} \left(\dfrac{2m^* k_B T}{\pi \hbar^2}\right)^{1/2} F_{-1/2}\left(\dfrac{-E_j - q\phi_c}{k_B T}\right) \tag{3.51}$$

where $n(r,\theta)$ is called the density of volume charge, g_v is the valley degeneracy, k_B denotes Boltzmann's constant, T is temperature, m^* is effective mass of electron, $\hbar$ is reduced Plank's constant, and $F_{-\frac{1}{2}}$ is Fermi Dirac integral of order $-\frac{1}{2}$. E_j is the energy in subband and ϕ_c is the central potential.

Since equation 3.49 and 3.50 are same, in order to equate them, first it is required to establish a relation between surface potential and central potential. Secondly, subband energy (E_j) must consider the effect of both geometrical and electrical confinement of charge carriers. Then it will make subband energy E_j implicitly dependent on potential which is called surface potential.

Here, Poisson's equation is used to establish a relation between surface potential and central potential. By using the cylindrical coordinate system in Poisson's equation for a nanowire of intrinsic semiconductor can be given as

$$\dfrac{1}{r^2}\dfrac{\partial^2 \phi}{\partial \theta^2} + \dfrac{1}{r}\dfrac{\partial}{\partial r}\left(r\dfrac{\partial \phi}{\partial r}\right) = \dfrac{q}{\varepsilon_s}(n(r,\theta)) \tag{3.52}$$

where the volume charge density $n(r,\theta)$ is given by

$$n(r, \theta) = \sum_{j=g_v}^{n} \left(\frac{2m^*k_BT}{\pi\hbar^2}\right)^{1/2} \times F_{-1/2}\left(\frac{-E_j - q\phi_c}{k_BT}\right) |\psi_j(r,\theta)|^2 \quad (3.53)$$

where $\psi_j(r, \theta)$ is called the wave function of the jth subband, which is expressed in terms of the Bessel function. The Bessel function makes the modeling complicated, especially while incorporating the electrical confinement. However, the higher quantum confinement and lower effective mass of semiconductor material leads to the significant volume inversion on nanowires. The approximation of wave function can be expressed as

$$|\psi_j(r, \theta)| = \frac{1}{\sqrt{\pi}R_o} \quad (3.54)$$

to satisfy the condition of normalization. Using CCDA, $n(r, \theta)$ can be written as

$$n(r, \theta) \approx \frac{1}{\pi R_o^2} \sum_{j=g_v}^{n} \left(\frac{2m^*k_BT}{\pi\hbar^2}\right)^{1/2} \times F_{-1/2}\left(\frac{-E_j - q\phi_c}{k_BT}\right) |\psi_j(r,\theta)|^2 \quad (3.55)$$

$$n(r, \theta) = -\frac{Q_s}{q\pi R_o^2} \quad (3.56)$$

Substituting equation (3.6) into (3.3) as a function of r can be written as

$$\phi(r) = \phi_c - \frac{Q_s}{q\pi R_o^2 \epsilon_s}\left[\frac{r^4}{4}\right] \quad (3.57)$$

Evaluating (3.7) at r = R0, a relation between surface potential and central potential can be written as follows:

$$\phi(r)|_{r=R_o} = \phi_c - \frac{Q_s}{q\pi\epsilon_s} \quad (3.58)$$

Using equation (3.8) and replacing it in equation (3.1) gives

$$Q_s = -C_{ins,eff}(V_G - \phi'_{ms} - \phi_c) \quad (3.59)$$

where $C_{ins,eff} = \frac{C_{ins}}{1+\frac{C_{ins}}{4\pi\epsilon_s}}$ is effective insulator capacitance, which captures the charge centroid.

3.5.6 Subband Energy Modeling

The bandgap energies and optical properties of nanowires have been calculated with various semi-spherical methods such as KP theory, pseudopotential, tight binding, and

CCDA without dealing with the self-energy problem. To understand the charge transport in semiconductor nanowires, the subband structure plays a key role. The most fundamental characteristics of nanowires i.e., if characteristics or conductance are directly related to the bandgap and subband energies (Chia et al., 2012).

The subband energy of semiconductor nanowires can be given as

$$E_j = E_{j,g} + \Delta E_j, e \qquad (3.60)$$

where $E_{j,g}$ refers to the subband energy due to geometrical confinement, and E_j,e refers to subband energy due to electrical confinement. The subband energy in electrical confinement can be modeled as

$$\Delta E_j, e = \langle \psi_j^* | \check{\phi} | \psi_j \rangle \qquad (3.61)$$

where $\check{\phi}$ is called perturbing potential.

$$\check{\phi}(r) = \phi(r) - \phi_c \qquad (3.62)$$

Here, ϕ_c is used as a reference to calculate E_j as specified in equation (3.2). Using the CCDA, equation (3.11) can be calculated analytically as

$$\Delta E_j, e = \frac{qQ_s}{8\pi\epsilon_s} \qquad (3.63)$$

The eigenvalue of subband energies can be derived by solving Schrödinger's equation in the cylindrical nanowire. The structure of the nanowire used for modeling subband energy is shown in the figure. Different values of effective mass are used for the effective mass of semiconductors and the effective mass of insulators. So the wave function allows penetrating the gate insulator. For the calculation, the following set of equations is derived:

$$\frac{\eta}{m_s^*} \frac{J_{l-1}(\eta) - J_{l+1}(\eta)}{J_l(\eta)} = -\frac{K_{l-1}(\zeta) + K_{l+1}(\zeta)}{m_{ins}^* K_l(\zeta)} \qquad (3.64)$$

$$\eta^2 + \frac{m_s^*}{m_{ins}^*}\zeta^2 = \frac{2m_s^* \Delta\phi R_o^2}{\hbar^2} \qquad (3.65)$$

where $J_l(x)$ is the first kind order and order l Bessel function, and $K_l(x)$ is the second kind and order l Bessel function. R_o is the radius of semiconductor nanowire, and $\eta = \gamma R_o$ and $\zeta = \alpha R_o$, where α and γ are related to the subband energy eigenvalues (E) as

$$\gamma^2 = \frac{m_s^*}{\hbar^2}E, \qquad (3.66)$$

$$\alpha^2 = \frac{2m_{ins}^*}{\hbar^2}(\Delta\phi - E) \tag{3.67}$$

By using the above set of equations, the subband energy eigen values can be obtained easily.

For $l = 0$

$$\zeta = -\frac{1}{2} + \frac{m_{ins}}{m_s}\eta\left[\frac{\cos(\eta - \phi_1)}{\cos(\eta - \phi_2)}\right] \tag{3.68}$$

For $l = \pm 1$

$$\zeta = -\frac{1}{2} - \frac{m_{ins}}{m_s}\left[\eta\frac{\cos(\eta - \phi_2)}{\cos(\eta - \phi_1)}\right] \tag{3.69}$$

where $\phi_1 = \left(\frac{\pi}{2}\right) + \left(\frac{\pi}{4}\right)$ and $\phi_2 = \left(\frac{\pi}{4}\right)$

It must be noted that the above solution for subband energies assumes a lead's potential profile.

3.6 HEALTHCARE APPLICATIONS OF NANOWIRE FETS

3.6.1 Detection of Viruses

Infections are the most widely recognized reason for human illness (Stadler et al., 2003); also, there is expanding worry about their ability to be used as an organic weapon (Atlas, 2003). The capacity to identify infections more adequately will prompt improved wellbeing and expanded security. Until now, Si nanowire sensors have effectively identified numerous perilous infections, including dengue (Zhang et al., 2010), influenza A H3N2 (Shen et al., 2012), H1N1 (Kao et al., 2011), and HIV (Inci et al., 2013). It is additionally conceivable to identify two particular infections at the single infection level (Patolsky et al., 2004). Antibodies specific to the objective infection are chosen and interact to the surface of nanowire FET sensor, trailed by the manufacturing of the nanowire FETs. When viruses are detected by the nanowire FET sensor, the conductivity of nanowires is affected by the antibodies. The flow of virus detection is shown in Figure 3.8.

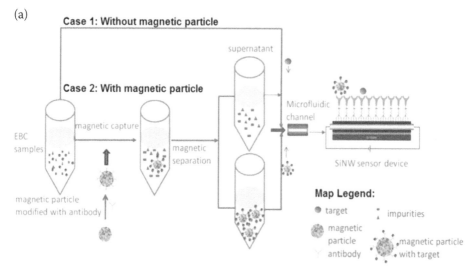

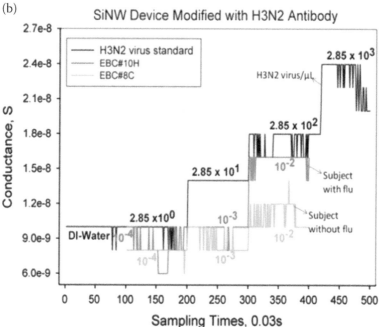

FIGURE 3.8 Detection flow of virus by nanowire FET sensor (Shen et al., 2012).

3.6.2 Detection of Biomarkers

Recent development in the research of genomics shows a number of biomarkers that have good properties to detect the infection (Etzioni et al., 2003). The heterogeneity of confounded infections, like malignancy, precludes the test of a single marker from giving sufficient results and requires the expanded requirements for a number of biomarkers (Wulfkuhle et al., 2003). The sensor with a p-type silicon nanowire was first used to electrical recognition of proteins in the solution (Cui et al., 2001). The detection process flow of biomarkers is shown (Figure 3.9).

In ongoing years, numerous examinations have zeroed in on building extensive stages for the constant discovery (Kim et al., 2007) or direct recognition of infections from blood (Stern et al., 2010). The sort of identified particles has likewise reached out to incorporate a wide assortment of biomarkers, for example, heart biomarkers (Chua et al., 2009). Shahad et al. recently built up a nanowire-based disease analysis framework utilizing breath volatolome as the information (Shehada et al., 2015).

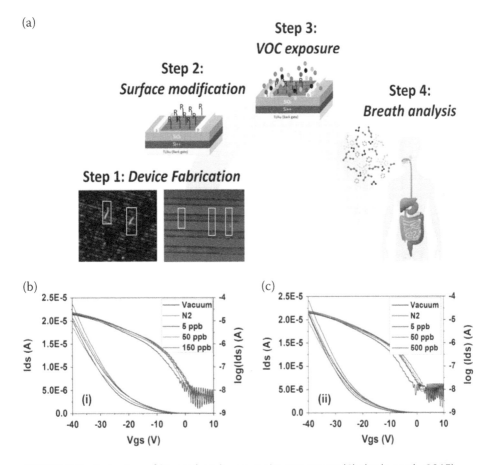

FIGURE 3.9 Detection of biomarkers by nanowire FET sensor (Shehada et al., 2015).

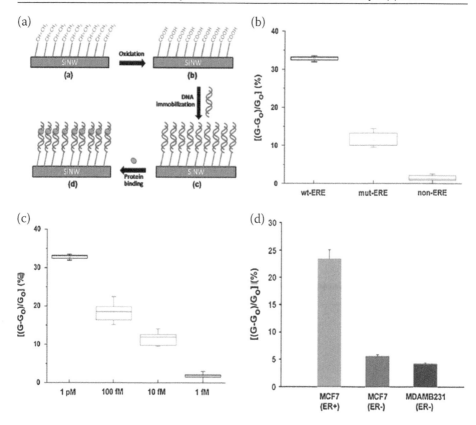

FIGURE 3.10 Detection of DNA and RNA by nanowire FET sensor (Zhang et al., 2011).

3.6.3 Detection of DNA and RNA

Nanowire FET sensors are fit for recognizing explicit arrangements of DNA (Zhang et al., 2008). Reciprocal single-abandoned groupings of PNAs are utilized as receptors of DNA on the surfaces of silicon nanowire FET sensors. Figure 3.10 shows the detection of interaction between DNA and protein.

3.6.4 Discovery of Medication

Nanowire FET-based biosensors have good characteristics, which make it possible to use nanowire FET biosensors in the discovery of medicines. Nanowire can be used as the drug carrier to treat the various diseases. Generation and measurement of electrophysiological signals is also done using nanowire FET biosensors.

As shown in Figure 3.11(a), without the nanowires, DOX atoms can just diffuse in flawless cell films, which bring about their fast expulsion from drug-safe malignancy cells and trouble in aggregating DOX particles in the phone cores.

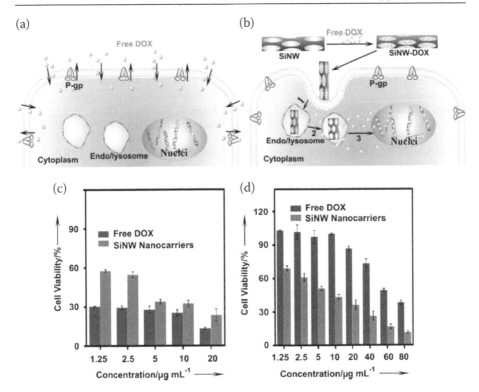

FIGURE 3.11 Mechanisms utilized to overcome the MDR effect of free DOX (Peng et al., 2014).

3.7 SUMMARY

In this chapter, the structure of advanced MOSFET has been discussed in brief. The modeling of nanoscale devices introduced ballistic transport and quantum transport of charge carriers. This chapter explains the various applications of nanowire FETs for healthcare applications such as biosensing. This chapter reviews the applications of nanowire FETs for virus detection, biomarker detection, DNA detection, and their applications for the discovery of drugs. In addition, it also describes the modeling of nanowires.

REFERENCES

Anantram, M. P., Lundstrom, M. S., & Nikonov, D. E. (2008). Modeling of Nanoscale Devices. *Proceedings of the IEEE*, 96(9).

Andrew, C. E., Chia & Pierre, R. R. L. (2012). Analytical model of surface depletion in GaAs nanowires. *Journal of Applied Physics*, *112*, 1–7.

Atlas, R. M. (2003). Bioterrorism and biodefence research: changing the focus of microbiology. *Nature Reviews Microbiology*, *1*(1), 70–74.

Bhattacharya, D., & Jha, N. K. (2014). FinFETs: From Devices to Architectures. *Advances in Electronics*.

Chang, L., Choi, Y.-K, Kedzierski, J., Lindert, N., Xuan, P., Bokor, J., Hu, C. & King, T. J. (2003). Moore's law lives on [CMOS transistors]. *IEEE Circuits Devices Magazine*. *19*(1), 35–42.

Chang, C., Ha, R., Xiong, B., Hu, & King (2003). Extremely scaled silicon nano-CMOS device. *Proceedings of the IEEE*, *91*(11), 1860–1873.

Chia, T., Li, Z., Mi & Comedi (2012). Electrical transport and optical model of GaAs-AlInP core-shell nanowires. *Journal of Applied Physics*, *111*.

Choi, A., Lindert, S., King, B., & Hu C. (2000). Ultrathin-body SOI MOSFET for deep-sub-tenth micron era. *IEEE Electron Device Letters*, 21, 254.

Chua, J. H., Chee, R. E., Agarwal, A., Wong, S. M., & Zhang, G. J. (2009). Label-free electrical detection of cardiac biomarker with complementary metal-oxide semiconductor-compatible silicon nanowire sensor arrays. *Analytical chemistry*, *81*(15), 6266–6271.

Cohen-Karni, T., Casanova, D., Cahoon, J. F., Qing, Q., Bell, D. C., & Lieber, C. M. (2012). Synthetically encoded ultrashort-channel nanowire transistors for fast, pointlike cellular signal detection. *Nano letters*, *12*(5), 2639–2644. (Cohen et al., 2012)

Cui, Y., Wei, Q., Park, H., & Lieber, C. M. (2001). Nanowire nanosensors for highly sensitive and selective detection of biological and chemical species. *Science*, *293*(5533), 1289–1292.

Datta (1996). Electronic Conduction in Mesoscopic Systems. Cambridge, U.K.: Cambridge University Press.

Doyle, B. S., Datta, S., Doczy, M., Hareland, S., Jin, B., Kavalieros, J., Linton, T., Murthy, A., Rios, R., & Chau, R. (2003). High performance fully-depleted tri-gate CMOS transistors. *IEEE Electron Device Letters*, *24*, 263–265.

Etzioni, R., Urban, N., Ramsey, S., McIntosh, M., Schwartz, S., Reid, B., ... & Hartwell, L. (2003). The case for early detection. *Nature reviews cancer*, *3*(4), 243–252.

Fiori G., & Iannaccone, G. (2007). Three-Dimensional Simulation of One-Dimensional Transport in Silicon Nanowire Transistors. *IEEE Transactions on Nanotechnology*, *6*, 524–529.

Ganeriwala, M. D., Ruiz, F. G., Marin, E. G., & Mohapatra, N. R. (2019). A Compact Charge and Surface Potential Model for III–V Cylindrical Nanowire Transistors. *IEEE Transactions on Electron Devices*, *66*(1).

Ganeriwala, M. D., Yadav, C., Ruiz, F. G., Marin, E. G., Chauhan, Y. S., & Mohapatra, N. R. (2017). Modeling of Quantum Confinement and Capacitance in III–V Gate-All-Around 1-D Transistors, *IEEE Transactions on Electron Devices*, *64*(12), 4889–4896.

Gnani, E., Reggiani, S., Gnudi, A. Parruccini, P., Colle, R., Rudan, M., & Baccarani, G. (2007). Band-Structure Effects in Ultrascaled Silicon Nanowires. *IEEE Transactions on Electron Devices*, *54*, 2243–2254.

Granzner, P., Schippel, & Schwierz (2014). Empirical Model for the Effective Electron Mobility in Silicon Nanowires. *IEEE Transactions on Electron Devices*, *61*.

Lundstrom (2000). Fundamentals of Carrier Transport, 2nd ed. Cambridge, U.K.: Cambridge Univ. Press.

Han, J. W., Kim, C. J., & Choi, Y. K. (2008). Universal potential model in tied and separated Double-Gate MOSFETs with consideration of symmetric and asymmetric structure. *IEEE Transactions on Electron Devices*, *55*(6), 1472–1479.

Inci, F., Tokel, O., Wang, S., Gurkan, U. A., Tasoglu, S., Kuritzkes, D. R., & Demirci, U. (2013). Nanoplasmonic quantitative detection of intact viruses from unprocessed whole blood. *ACS nano*, *7*(6), 4733–4745

Industry Association (2004). International Technology Roadmap for Semiconductors Update.

Jain, A., Sharma, S. K., & Raj, B. (2016). Design and analysis of high sensitivity photosensor using Cylindrical Surrounding Gate MOSFET for low power applications. *Engineering Science and Technology, an International Journal*, 19(4), 1864–1870.

Jain, A., Sharma, S., & Raj, B. (2018). "Analysis of Triple Metal Surrounding Gate (TM-SG) III–V Nanowire MOSFET for Photosensing Application," *Opto-electronics Journal*, Elsevier, 26(2), 141–148, May.

Jin, S., Tang, T. W., & Fischetti, M. V. (2008). Simulation of Silicon Nanowire Transistors Using Boltzmann Transport Equation Under Relaxation Time Approximation. IEEE Transactions on Electron Devices, 55(3), 727–736.

Kao, L. T. H., Shankar, L., Kang, T. G., Zhang, G., Tay, G. K. I., Rafei, S. R. M., & Lee, C. W. H. (2011). Multiplexed detection and differentiation of the DNA strains for influenza A (H1N1 2009) using a silicon-based microfluidic system. *Biosensors and Bioelectronics*, 26(5), 2006–2011.

Khan, F. S., Davies, J. H., & Wilkins, J. W. (1987). Quantum transport equations for high electric fields. *Physical Review B*, 36(5), 2578.

Kim, A., Ah, C. S., Yu, H. Y., Yang, J. H., Baek, I. B., Ahn, C. G., ... & Lee, S. (2007). Ultrasensitive, label-free, and real-time immunodetection using silicon field-effect transistors. *Applied Physics Letters*, 91(10), 103901.

Lee, Y., Kakushima, K., Natori, K., & Iwai, H. (2015). Modeling of quasi-ballistic transport in nanowire metal-oxide-semiconductor fieldeffect transistors. *Journal of Applied Physics*.

Lilienfeld J., J. E. (1930). Method and apparatus for controlling electric currents, U.S. Patent 1,745,175, issued January 28, 1930.

Majima, H., Ishikuro, H., & Hiramoto, T. (2000). Experimental evidence for quantum mechanical narrow channel effect in ultra-narrow MOSFET's. *IEEE Electron Device Letters*, 21, 396–398.

Park, J. T., Colinge, J. P., & Diaz, C. H. (2001). Pigate SOI MOSFET, *IEEE Electron Device Letters*, 22, 405–406.

Patolsky, F., Zheng, G., Hayden, O., Lakadamyali, M., Zhuang, X., & Lieber, C. M. (2004). Electrical detection of single viruses. *Proceedings of the National Academy of Sciences*, 101(39), 14017–14022.

Paul, B. C., Tu, R., Fujita, S., Okajima, M., Lee, T. H., & Nishi, Y. (2007). An Analytical Compact Circuit Model for Nanowire FET. *IEEE Transactions on Electron Devices*, 54, 1637–1644.

Peng, F., Su, Y., Ji, X., Zhong, Y., Wei, X., & He, Y. (2014). Doxorubicin-loaded silicon nanowires for the treatment of drug-resistant cancer cells. *Biomaterials*, 35(19), 5188–5195.

Ramayya, E. B., Vasileska, D., Goodnick, S. M., & Knezevic, I. (2007). Electron Mobility in Silicon Nanowires. *IEEE Transactions on Nanotechnology*, 6, 113–117.

Reza, S., Bosman, G., Islam, M. S., Kamins, T. I., Sharma, S., & Williams, R. S. (2006). Noise in Silicon Nanowires. *IEEE Transactions on Nanotechnology*, 5, 523–529.

Roosbroeck (1953). The transport of added current carriers in a homogeneous semicon- ductor. *Physical Review*, 91(2), 282–289.

Sharma, S. K., Jain, A., & Raj, B. (2018). Analysis of triple metal surrounding gate (TM-SG) III–V nanowire MOSFET for photosensing application. *Opto-Electronics Review*, 141–148.

Singh, A., Khosla, M., & Raj, B. (2016). Comparative Analysis of Carbon Nanotube Field Effect Transistor and Nanowire Transistor for Low Power Circuit Design. *Journal of Nanoelectronics and Optoelectronics*, 11, 1–6.

Shafizade, D., Shalchian, M., & Jazaeri, F. (2019). Ultrathin Junctionless Nanowire FET Model, Including 2-D Quantum Confinements. *IEEE Transactions on Electron Devices*, 66(9), 4101–4106.

Shehada, N., Brönstrup, G., Funka, K., Christiansen, S., Leja, M., & Haick, H. (2015). Ultrasensitive silicon nanowire for real-world gas sensing: noninvasive diagnosis of cancer from breath volatolome *Nano Letters, 15*(2), 1288–1295.

Shen, F., Wang, J., Xu, Z., Wu, Y., Chen, Q., Li, X., Jie, X., Li, L., Yao, M., Guo, X., & Zhu, T. (2012). Rapid Flu Diagnosis Using Silicon Nanowire Sensor. *Nano Letters, 12*(7), 3722–3730.

Shin, M. (2007). Quantum Simulation of Device Characteristics of Silicon Nanowire FETs. *IEEE Transactions on Nanotechnology, 6*, 230–237.

Shockley, W. (1950). Electrons and Holes in Semiconductors. Van Nostrand: New York.

Stadele (2002). Influence of source-drain tunneling on the subthreshold behavior of sub-10-nm double-gate MOSFETs. *In Proceedings of European Solid-State Device Research Conference*. Firenze, Italy, 135–138.

Stadler, K., Masignani, V., Eickmann, M., Becker, S., Abrignani, S., Klenk, H.-D., & Rappuoli, R. (2003). SARS—beginning to understand a new virus. *Nature Reviews Microbiology, 1*(3), 209–218.

Stern, E., Vacic, A., Rajan, N. K., Criscione, J. M., Park, J., Ilic, B. R., Mooney, D. J., Reed, M. A., & Fahmy, T. M. (2010). Label-free biomarker detection from whole blood. *Nature Nanotechnology, 5*(2), 138–142.

Subramaniam, S., Awale, R. N., & Joshi, S. M. (2013). Drain current models for single-gate MOSFETs & undoped symmetric & asymmetric double-gate SOI MOSFETs and quantum mechanical effects: a review. *International Journal of Engineering Science and Technology, 5*, 96–105

Wang, J., & Lundstrom, M. (2002). Does source-to-drain tunneling limit the ultimate scaling of MOSFETs? *IEDM Technical Digest*: San Francisco, 707–710.

Wang, J., Rahman, A., Ghosh, A., Klimeck, G., & Lundstrom, M. (2005). On the Validity of the Parabolic Effective-Mass Approximation for the I–V Calculation of Silicon Nanowire Transistors. *IEEE Transactions on Electron Devices, 52*, 7.

Wulfkuhle, J. D., Liotta, L. A., & Petricoin, E. F. (2003). Proteomic applications for the early detection of cancer. *Nature reviews cancer, 3*(4), 267–275.

Yang, F. L., Chen, H. Y., Chen, F. C., Huang, C. C., Chang, C. Y., Chiu, H. K., Lee, C. C., Chen, C. C., Huang, H. T., Chen, C. J., Tao, H. J., Yeo, Y. C., Liang, M. S., & Hu, C. (2002). 25 nm CMOS Omega FETs. *Proceedings of IEEE Electron Devices Meeting*, San Francisco, CA, USA, 255–258.

Yesayan, A., Jazaeri, F., & Sallese, J.-M. (2020). Analytical Modeling of Double-Gate and Nanowire Junctionless ISFETs. *IEEE Transactions on Electron Devices, 67*(3), 1157–1164.

Zhang, G. J., Huang, M. J., Ang, J. J., Liu, E. T., & Desai, K. V. (2011). Self-assembled monolayer-assisted silicon nanowire biosensor for detection of protein-DNA interactions in nuclear extracts from breast cancer cell *Biosensors and Bioelectronics, 26*(7), 3233–3239.

Zhang, G. J., Zhang, G., Chua, J. H., Chee, R. E., Wong, E. H., Agarwal, A., Buddharaju, K.D., Singh, N., Gao, Z., & Balasubramanian, N. (2008). DNA sensing by silicon nanowire: charge layer distance dependence *Nano Letters, 8*(4), 1066–1070.

Zhang, G. J., Zhang, L., Huang, M. J., Luo, Z. H. H., Tay, G. K. I., Lim, E. J. A., Kang, T. G., & Chen, Y. (2010). Silicon nanowire biosensor for highly sensitive and rapid detection of Dengue virus. *Sensors and Actuators B: Chemical, 146*(1), 138–144.

Zhao, H., Kim, R., Paul, A., Luisier, M., Klimeck, G., Ma, F. J., Rustagi, S. C., Samudra, G. S., Singh, N. & Lo, G. Q., Kwong, D. L. (2009). Characterization and Modeling of Subfemtofarad Nanowire Capacitance Using the CBCM Technique. *IEEE Electron Device Letters, 30*(5), 526–528.

TFET-Based Sensor Design for Healthcare Applications

4

Tulika Chawla and Mamta Khosla
Nanoelectronics Reasearch Lab, Department of Electronics and Communication Engineering, National Institute of Technology Jalandhar, India

Balwinder Raj
Department of Electronics and Communication Engineering, NITTTR Chandigarh, India

Contents

4.1	Introduction	70
4.2	Types of Sensors and Their Applications in the Healthcare Sector	71
4.3	TFET Device Operation and Working Principle	75
4.4	Structures and Workings of TFET-Based Biosensors	77
4.5	Different Architectures of TFET-Based Biosensors	78
	4.5.1 Doped Dielectric Modulated TFET-Based Biosensor	79
	4.5.2 Charge Plasma-Based Junctionless TFET Biosensors	83
4.6	Structures and Workings of TFET-Based Gas Sensors	85
4.7	Different Architectures of TFET-Based Gas Sensors	89
	4.7.1 Metallic Gate Doped TFET-Based Gas Sensors	89
	4.7.2 Metallic Gate-Based Charge Plasma Based TFET Gas Sensors	90
	4.7.3 Conducting Polymer (CP)-Based GAA-TFET Gas Sensors	92

4.8	Structures and Workings of TFET-Based Photosensor	92
4.9	Different Architectures of TFET-Based Photosensors	94
	4.9.1 Hybrid Photosensors	94
	4.9.2 Hetero-Material-Based Charge Plasma-Based TFET Photosensors	95
4.10	Summary	96
References		97

4.1 INTRODUCTION

In the field of medical health care, there has been a growth in demand for gadgets based on nano-sensors that are user-friendly and have a quick reaction time for diagnosing various ailments in the human population. By creating new medical gadgets with novel technologies integrated with various nano-sensors, the healthcare industry is moving toward improving society's health condition. Medical healthcare equipment's major goal is to be able to evaluate health state, illness onset, and sequencing. Early detection of syndromes is critical for patient survival and accurate illness prognosis; therefore, biosensors must have great sensitivity and a practical structure. Sensor-based medical devices that are implanted as well as wearables have become a reality due to advancements in medical technology. These devices are becoming more common in society over time. The primary issue is to create low-noise, low-power, low-area nano-sensor devices while keeping patient safety in mind. A gas sensor can also be used to identify illnesses. Changes in the quantities of the trace species can indicate the presence of a disease. The volatile organic molecule acetone can be used to distinguish between healthy and diabetic people in the case of diabetes [1]. Pulse oximetry, which measures the amount of oxygen in the blood (see "Optical Sensors in Pulse Oximetry"), heart-rate monitors, blood diagnostics such as blood glucose monitoring, urine analysis, and dental color matching are all examples of current medical applications that use optoelectronic sensors. Pulse oximetry is a cutting-edge non-invasive method for detecting the proportion of hemoglobin (Hb) saturated with oxygen, which is crucial for providing anesthesia, assessing the function of the respiratory system, or diagnosing diseases including pneumonia, pleurisy, and asthma [2]. Sensor technology offers a lot of potential in the medical business; for example, sensors in cancer therapy, non-invasive detection, and other areas have shown a lot of promise. Furthermore, as the threat of bioterrorism grows, devices embedded with different sensors are needed that can identify contaminated bioagents in the atmosphere rapidly, reliably, and correctly is critical in order to control the infection through virus and treatment [3].

Sensor technology will continue to advance in the future, and sensors in the area of contemporary medicine may become more widespread, promoting the advancement of modern medical diagnostic and treatment techniques. In many ongoing years, a variety of FET-based nano-sensors have been created, including ion sensitive field-effect transistors (ISFETs), dielectric modulated field effect transistors (DM-FETs), and tunnel field effect transistors (TFETs) [4–6]. Although FET-based sensors have a number of advantages over other types of sensors, the existence of short channel effects and a significant subthreshold swing issue in the FET restricts device sensitivity and increases power consumption due to thermionic electron emission. Conventional FET sensors have major limits as far as highest sensitivity and lowest detection time. TFET-based nano-sensors, on the other hand, can overcome these drawbacks by employing a fundamentally different current injection process known as band-to-band tunnelling to obtain better sensitivity and shorter reaction times. Although, there has been significant progress in the development of TFET-based nano-sensors, complete information on different structure-based TFET nano-sensors and their application is not available in literature at this instance. In this chapter, we offer a comprehensive overview of TFET-based nano-sensors along with their applications. This chapter is organized into 10 sections. Section 4.2 explains the types of sensors and their applications in the healthcare sector. Section 4.3 discusses the TFET device operation and working principle. Section 4.4 discusses the structures and workings of TFET-based biosensors and Section 4.5 different architectures of TFET-based biosensors are discussed. Section 4.6 discusses the structures and workings of TFET-based gas sensors and different architectures of TFET-based gas-sensors are discussed in Section 4.7. Section 4.8 discusses the structures and workings of TFET-based photosensors and the different architectures of TFET-based photosensors are explained in the Section 4.9. A summary of this chapter is discussed in Section 4.10.

4.2 TYPES OF SENSORS AND THEIR APPLICATIONS IN THE HEALTHCARE SECTOR

Biosensor history began in 1962 when the enzyme electrodes were developed by scientist Leland C. Clark [3]. After that, research groups from different arenas, such as physics, chemistry, integration (VLSI), and material science, gathered to develop more sophisticated and stable biosensing equipments. Applications for these equipments are in the area of medical, farming, and biotechnology areas as well as in the defence area in addition to detecting and preventing bioterrorism. The biosensors can be of numerous sorts depending on the transduction method employed, i.e., resonant biosensors, optical biosensors, thermal sensor biosensors, ion sensitive field effect transistors (ISFET) biosensors, and electrochemical biosensors. The first ion sensitive FET-based biosensor was investigated in 1970 by Befgveld [7]. The biosensor based on the ion sensitive field effect (ISFET) is increasingly used for the recognition of charged biomolecules. Although ISFET is not able to detect uncharged biomolecules, therefore ISFETs are superseded with dielectric-modulated field effect transistor-based biosensors. Biosensors can be further classified on the basis of detection techniques: (a) Label (marker)-based biosensors and (b) label-free biosensors. In label-based detection, the word "label" is an identifier for any chemical or temporary external molecule that increases the binding sites to be used to sense the intended analysis. Electrochemical sensors can easily detect the labels [8,9]. The biosensor electrochemically recognizes many biomolecules in the human body, such as protein, biotin, uricase, DNA, glucose, and hemoglobin, etc. Even if label-based techniques are efficient and have large detection limits, their use may lead to difficulty in multiplexing and altering the inherent characteristics of target molecules. A wrong label, for instance, may lead to the creation of a medicine with side effects. In contrast, label-free detection methods use molecular biophysical characteristics that includes the molecular weight, the refractive indices, and the molecular charge to monitor molecule presence and its activity. In addition, molecular events may be monitored in real time with these techniques. Molecular interactions are transduced into mechanical, electric, or optical signals in a typical biosensing process and hence observable without label probes. Label-free technology is better and is more advanced than label-based technology [10–12]. Figure 4.1 shows some of the advantages of the label-free biosensors. In addition to these advantages, this technique is a cheaper technique than label-driven ones. Consequently, the label-free method of biosensing became a systemic and universal approach at nano-scale testing that is extremely adaptable for remote and field diagnosis.

Label-free detection accuracy makes it suitable for biosensing applications; label-free processes compete for improved stability and a high signal to noise ratio (SNR). In recent times, biosensors based on FET are taken into consideration due to their advantages in label-free electrical detection, miniaturized size and weight, low-priced mass production, and the potential to integrate sensors and measuring systems in a single-chip environment. Diversified use of biosensors can be visualized in different fields through Figure 4.2 [3]. Blood glucose sensing is an important application due to its vast market potential. However, biosensors offer significant market potential in other application sectors, such food and beverage biosensor-based devices, environment sampling, and non-invasive clinical diagnostic tools.

A mechanism that can identify illnesses is long recognized as the human breath. Gas sensors are now capable of diagnosing a wide spectrum of illnesses through human

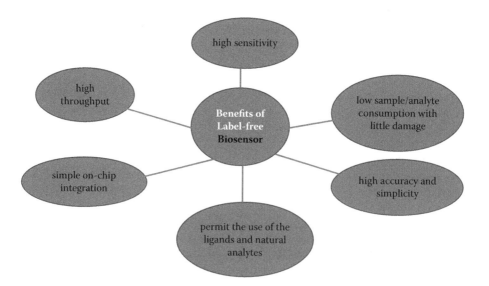

FIGURE 4.1 Advantages of the label-free biosensors.

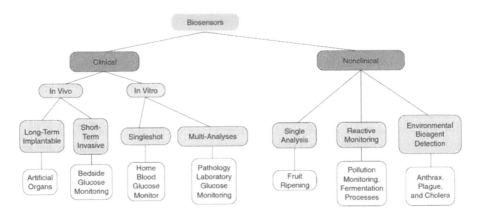

FIGURE 4.2 Possible applications of biosensors.

breath using advances in contemporary nanotechnology [13]. Some diseases like cancer to diabetes are required to be treated from the initial stage of the illness for both increasing the patient results and lowering the cost of therapy. Thus, fast and non-invasive human breath gas sensors become a suitable choice over the existing way that will meet the requirement [14]. The working mechanism of TFET-based gas sensor is presented in Section 4.6, elaborating on the different methods to increase the selectivity, and sensitivity of the gas sensor that were previously reported in various researches.

Gas sensors for many sectors, from agriculture to environmental monitoring, are utilized for several purposes. The market, which is expected to develop worldwide to more than $15 billion by 2022 [15], is among those that the biomedical sensor sector is most promising to make use of gas sensor advantages. Gas sensors particularly developed for human breath analysis are a reliable option for a fast and accurate detection of illnesses compared to existing conventional diagnosis procedures such as slow-through and intrusive blood tests. This capability for gas sensors in the detection of diseases is mainly due to the nature of human breath which also consists of predictable gases like N_2, O_2, H_2O vapors, and CO_2 and also comprised of additional trace species, such as NH_3, C_3H_6O, CH_3CH_2OH, C_2H_4O, and C_3H_8O and others at concentrations that are as small as parts per trillion (ppt) [16,17]. When these species' quantities vary, this will indicate the sign of the certain ailment for example: volatile organic molecule acetone might function as a marker between healthy people and diabetic patients in case of diabetes [18,19]. Similarly, the shift in H_2S, NH_3, NO, and toluene trace species may be utilized to detect halitosis, renal dysfunction, asthma, and lung cancer [20–22] correspondingly. Therefore, gas sensors need to be very sensitive for the purpose of an efficient biomedical diagnosis instrument. The quantity of these biomarkers in the exhaled respiration, however, may be modified by several characteristics such as ambient circumstances or the medical history of patients. In reality, the biomarkers utilized should be taken into account from patient to patient. In the respiration of smokers and former smokers, toluene, for example, is already detected in high amounts [23]. However, afterward applying normalisation, substantial dissimilarities may be reached between cancer individuals, smokers, and non-smokers [20].

Numerous kinds of gas sensors have thus been continually investigated and developed. Resistive [24,25], optical [26], electrochemical [27] and field-effect gas transistor sensor type (FET type) [28,29]; these four types of gas sensors are included in the context. One of the most well-researched types of gas sensors is resistive gas sensors. Recognition of targeted gases are done by evaluating the variation in sensors' resistance after exposure to the gases. It's inexpensive and simple to manufacture. Different sensing materials for example semiconductor metal oxides, TMDCs [30], carbon nanotubes [31,32], and graphene [33] etc. can also be used with a very basic structure. In addition, several researches have been done on the structure and chemical functionalization of sensing materials used for resistive gas sensors and chemical functionalisation, to increase sensitivity to the target gases. However, compared with other semiconductor sensors [34], they need to be big in size in order to decrease output fluctuation and achieve sufficient operating current. In addition, because of its huge sensing surface, a big heater is necessary to increase the working temperature and therefore significant consumption of power takes place. In general, optical gas sensors are comprised of three main portions: light source zone, a gas chamber, and a light detector. Each gas has distinct absorption properties at a particular light source wavelength. Therefore, the targeted gas molecules absorb a light of particular wavelength from the light source, and a light detector may detect and quantify the presence of target gas [35]. Because the working principle solely relates to the physical quality of the targeted gas and there is no chemical reaction between the target gas molecules and the optical sensor, optical gas sensors provide a better reliability because of their small output drift and their long life span. They are also highly selective and rapidly responding [36]. But they suffer in terms of miniaturization, poor portability, and expense, also, so their marketing is restricted and more improvements are needed. Ambient light interference also affects them [37]. Electrochemical gas sensors are composed of a working electrode, counter electrode, and an electrolyte solution that immerses both electrodes. By detecting the variation in current or electric potential of the two electrodes created by a reduction or oxidation process on the working electrode, electrochemical sensors sense targeted gases [27]. They can detect low concentrations of target gases and may be used at low power. They are inexpensive as well as they are not much influenced by environmental moisture. However, electrochemical sensors have a restricted temperature range as they may be used to dry out the electrolyte solution at largetemperature or small humidity. These sensors have a short life span and required the electrolyte solution to be maintained and calibrated [37]. FET-based gas sensor employing gate or channel as a sensing material, modulation in threshold voltage or drive current of those sensors happens when they are exposed to the target gas. In contrast to the three previously mentioned types of gas sensors, FET-type gas sensors are compatible with CMOS circuits. CMOS circuits can regulate the change in output signals and calibrate environmental changes during sensing such as moisture, temperature, etc. Therefore, FET gas sensors may be efficiently employed for the implementation of realistic highly accurate gas sensing systems, which can be simply built into CMOS circuits on a chip [38]. They can also be fabricated with the help of CMOS fabrication technology in extremely small dimensions at a reasonable cost. The FET gas sensors are of several kinds; for example, thin film transistor [39], FET catalytic gate metal, suspended gate FET [40], capacitively coupled FET [29,41] and horizontal floating-gate FET [42]. TFT gas sensors generally consiss of three regions: gate, source, and drain region that use a channel as an identifying material for an active layer. TFT sensors can sense target gases when the target gases are exposed to the sensors by measuring the modulated device parameter resulting from charge transmission or redistribution in the active layer.

At the present time, the demand for small-power and highly accurate photosensors/photodetectors is growing at a surprising rate in numerous applications; for example, flame sensors, chemical composition investigation, missile smoke estimation, medical analysis and treatment, scrutiny camera,s etc. [43] and also photosensitive interconnections for the communication of data in interchips [44] and photosensitive storage media [8]. FETs' compatibility with ICs has aided in the effective growth of these devices as photodetectors. The majority of these applications typically use silicon (Si)-based photodetectors because they need small volume, have a decent SNR, plentifully accessible, and are compatible with microelectronics. They are widely employed in defense and civil application fields. For sensors, sensitivity is an important parameter and this also affects the accuracy of the sensor. Further, to increase the sensitivity and accuracy of the nano-sensors, a band to band tunneling carrier injection mechanism based TFET device is employed for the sensing application in the different domains. Before discussing the working of TFET-based biosensor, gas sensor and photosensor, the fundamental basic principle needs to be understood, which is utilized in each sensor device.

4.3 TFET DEVICE OPERATION AND WORKING PRINCIPLE

TFETs are principally gated p-i-n devices in which band to band tunneling (BTBT) takes place parallel to oxide-Si interface. In this device, the tunneling takes place between the intrinsic and p+ regions. The basic structure of the conventional TFET is shown in Figure 4.3 [45].

In the absence of gate bias, the device is in the off-state as the energy barrier between the intrinsic area and the p+ area is much wider, as shown in the Figure 4.4. When positive gate bias increases (on-state), the energy bands in the intrinsic area start to move downward and the tunneling barrier becomes narrow, permitting the tunneling current to flow, as shown in Figure 4.4 [46]. Tunnelling probability is defined using the Wentzel Kramers-Brillouin (WKB) approximation [47]:

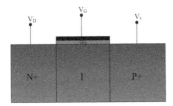

FIGURE 4.3 Structure of conventional TFET.

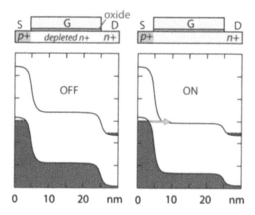

FIGURE 4.4 Energy-band diagram of TFET in OFF state and in ON state.

$$T(E) = e^{\left(\frac{-4\lambda\sqrt{(2m^*)}E_g^{3/2}}{3|e|\hbar(E_g+\Delta\Phi)}\right)} \tag{4.1}$$

where E_g, m^*, $\Delta\Phi$, and λ are the bandgap energy, effective carrier mass, energy gap between the conduction band of channel, and valence band of source and screening length, respectively. Here, screening length can be expressed as $\lambda = \sqrt{\frac{\varepsilon_{si}}{\varepsilon_{ox}}t_{ox}t_{si}}$, where λ can be decreased via high-k gate dielectric material or via thin gate oxide, which can enhance the tunneling probability. The value of tunneling current of the device [47] can be computed from equation (4.2):

$$I \propto e^{\left(\frac{-4\lambda\sqrt{(2m^*)}E_g^{3/2}}{3|e|\hbar(E_g+\Delta\Phi)}\right)} \cdot (\Delta\Phi) \tag{4.2}$$

Thus, from equation (4.1) and equation (4.2), it is clear that the tunneling probability and current depend upon the screening length, gate oxide thickness, dielectric constant, bandgap of the material, and effective carrier mass of the material.

Although TFET can attain low off-state current and subthreshold swing (SS) below 60 mV/dec due to BTBT mechanism, conventional Si-TFET cannot be used as a commercial device due to small on-state current and low I_{on}/I_{off} ratio. Various researchers have reported in the literature demonstrating the improvement in these performance parameters through device engineering [48–50]. It is further possible to reduce the SS of the conventional Si-TFET and thereby improve the performance through material, structure, and dielectric engineering. Hetero-structures [48–50] with small effective mass materials are used in source areas. Material engineering is implemented with a direct bandgap III-V compound semiconductor [51] such as GaSb and InGaSb, etc. to enhance the I_{on} current. A heterojunction structure [48–50] can achieve high I_{on}/I_{off} ratio. High-k dielectric materials are also used to improve the on-state current. Additionally, the point and average SS improves due to improved gate to channel coupling provided by a high-κ dielectric material. In addition to this, structural engineering is implemented to further improve the I_{on} current and I_{on}/I_{off} ratio. Various shapes of TFET such as U-shape TFET [52], symmetric U-shape TFET [53], L-shape TFET [54,55], and symmetric TFET [56,57] are credited to enhance the performance parameters. Further, pocket doping [58,59] is used in structures to make a device with better performance characteristics. Line tunneling is implemented to reduce the SS_{avg} and increase I_{on} current [49,60–62].

4.4 STRUCTURES AND WORKINGS OF TFET-BASED BIOSENSORS

The applications of biosensors are increasing quickly in the area of medical science. In clinical applications for diagnoses of diabetes mellitus, glucose biosensors are extensively employed, which need to be closely controlled on the levels of blood glucose. Eighty-five percent of the huge international market is reported for blood glucose biosensors used in homes. In primary-stage investigation of human interleukin (IL)-10.39, a new biosensor HfO_2 based biosensor has been employed [63]. In recent times, biosensors based on FET have a considerable demand due to their advantages in label-free electrical detection, the miniaturized size and small weight, affordable mass production, and the potential to integrate sensors and measuring systems in a single-chip environment. The basic principle of the electrical sensing through FET biosensors is dependent on the gating effect on the charged molecules, which can be directly observed by variation in electrical characteristics such as current and conductance, etc [64]. The dielectric coating of the semiconductor is operated using particular receptors for the detection of the required target biomolecules. Sensitivity is a key metric in the measurement of biosensor performance. For biomolecular detection at low concentrations and reduction time, an improved sensitivity is needed. Therefore, to increase the sensitivity, research starts on a TFET-based biosensor that is based on BTBT carrier injection method. The TFET-based biosensor consists of three terminals: gate, drain, and source, and the area channel is fitted with a biorecognition component. The schematic of a TFET-based nanowire biosensor is shown in Figure 4.5 [64]. This biorecognition component interacts with the targeted biomolecules and senses their existence by detecting the change in electrical activity. Afterward, the biological information is transformed directly into a computable signal. In other words, we can also define the operation of the TFET-based biosensors as a variation in the charge concentration at the channel's surface, which affects the modulation in effective gate voltage and this further leads to an increase in drain current due to narrowing of the tunneling length as a result of the gate effect. To measure the performance of biosensors, drain current sensitivity [65] is an important parameter. The formula for drain current sensitivity is

$$S_{Id}\% = \frac{(I_d^{bio} - I_d)}{I_d} * 100 \qquad (4.3)$$

78 Advanced Circuits and Systems for Healthcare and Security Applications

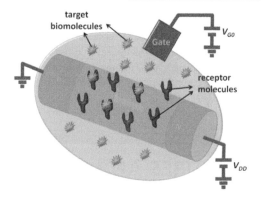

FIGURE 4.5 Schematic of a NW-TFET biosensor.

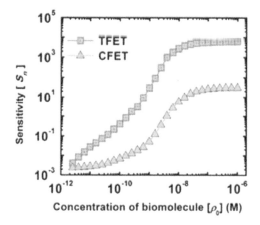

FIGURE 4.6 Comparative plot of sensitivity of TFET and CFET.

Here, I_d^{bio} and I_d are the values of drain current when the cavity is filled with dielectric material and the cavity is empty, respectively. Figure 4.6 [64] shows that TFET-based biosensors have high sensitivity compared to conventional FET biosensors. Further, in the following sub-section, different architectures of TFET-based biosensors are discussed that were developed by the various researchers to enhance the sensitivity of the sensors.

4.5 DIFFERENT ARCHITECTURES OF TFET-BASED BIOSENSORS

Sensitivity is an important parameter of the sensor, as sensitivity further affects the accuracy of the device, so research is done on various aspects of sensors based on sensor design and materials used for different regions etc. in order to further improve the sensitivity of the TFET-based sensor.

4.5.1 Doped Dielectric Modulated TFET-Based Biosensor

Label-free detection accuracy makes it suitable for biosensing applications; label-free processes compete for improved stability and a large SNR. Initially a MOSFET-based dielectric modulation (DM)-based biosensor is designed [66]. The fundamental principle behind the workings of DM-FET is that the cavity is etched in gate insulator material for the immobilization of the biomolecules within; due to this modulation in effective coupling of gate and channel occurs. The variations in electrical characteristics of the FET owing to a biomolecule's presence, absence, or characteristics (e.g., its charge) is subsequently utilized for sensitivity measurement in terms of V_{th} and I_d and for the aim of label-free detection of neutral and charged biomolecules. Conventional FET-based biosensors are replaced by TFET-based biosensors in order to overcome the limitation of conventional FET-based biosensors such as subthreshold swing, power consumption, and response time [64]. Combining the benefits of dielectric modulation with TFET-based biosensors led to the production of the development of DM-TFET-based biosensors. A dielectric modulated TFET biosensor has been described in [67] and is shown in Figure 4.7. The architecture has a hollow region that is to be occupied and immobilized by specific biomolecules. In the oxide layer, under the gate electrode, the cavity area was produced. After the biomolecules are stable, the device undergoes changes in the dielectric constant in the oxide due to their dielectric value. There is modulation in effective coupling between the gate and the oxide layer due to the modulation in the dielectric constant that leads to band bending (tunneling) that occurs in the channel region.

Several semiconductor devices have been utilized as sensors for biomedical applications; for example, temperature, light, and gas sensors are widely utilized in the

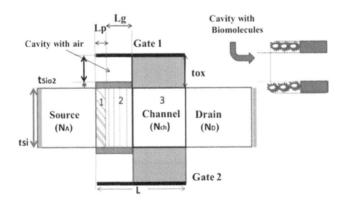

FIGURE 4.7 Architecture of p-n-p-n TFET dielectric modulated biosensor.

physiological measuring systems [4]. The main challenge is that the sensors are needs to be highly sensitive to changes in observable biomolecules for monitoring of health conditions for some specific ailment treatment. In order to enhance the sensitivity of biosensors, some structural modifications and material modifications are done in the earlier reported TFET biosensor devices. The below discussion gives you a glimpse how advancement in state-of-the-art TFET-based biosensor have taken place until now. In this paper [68], authors investigated the influence of the gate length on the sensitivity of the sensor. In the short gate-TFET design shown in Figure 4.8(b) [68], the decrease in the length of the gate increases its drain control over the band to band tunnelling procedure and this has been utilized for the sensing. Gate and drain voltages have dominant effects on the improvement in the sensitivity of short-gate biosensors. The gate and drain voltages are recognized as important parameter to optimize efficiency. Figure 4.8(a) and Figure 4.8(b) show the schematic of full-gate, double-gate, and short-gate double-gate TFET biosensor [68]. Figure 4.9 shows the comparative drain current sensitivity plot with respect to gate voltage, respectively [68]. The author proposed a dielectric modulated gate on drain TFET-based biosensors, shown in

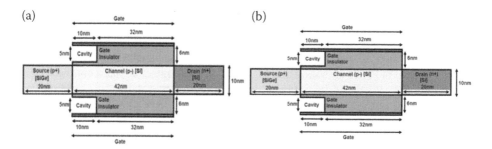

FIGURE 4.8 (a) Full-gate double-gate TFET biosensor; (b) short-gate double-gate TFET biosensor.

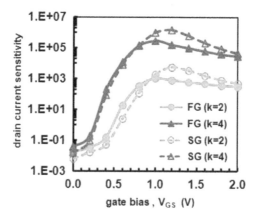

FIGURE 4.9 Comparative drain current sensitivity plot of FG and SG-DGTFET biosensor.

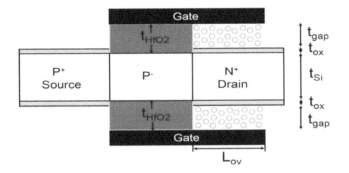

FIGURE 4.10 Structure of nanogap embedded gate on drain TFET biosensor.

Figure 4.10 [69]. In this reported device [69], the author uses an ambipolar current as a sensing parameter. Change in ambipolar current (I_{amb}) takes place with the change in biomolecule having different dielectric constants present in the nanocavity. Sensitivity achieved by the sensor lies in the range of 10^4 to 10^{10} as the permittivity of dielectric varies from K = 5 to K = 10. This biosensor not only provides the improved sensitivity but also takes the benefit of ambipolar current. Sensitivity for the proposed device is given by the relation of I_{amb} when there is no biomolecule and there is a biomolecule. A planar TFET biosensors suffers from several consequences due to manufacturing complexity and weak controllability of gate electrode over the channel. With the aim to solve the problems of the planar TFET biosensor, the author [65] proposed a biosensor using nanowire TFET (NW TFET), which improves the biosensor sensitivity and response time. Figure 4.11 [65] shows the different schematic view of a conventional nanowire (CNW) and source extended (SE) nanowire TFET biosensor. In this regard, a new design of cavity is created that is extended towards the source region from the etched portion under the gate. This modification in the cavity design increases the selectivity and sensing speed of the sensor. Figure 4.12 and Figure 4.13 show the drain current sensitivity plot for a CNW TFET biosensor and SE NWTFET biosensor, respectively [65]. The sensitivity of the device of the proposed work [65] is in the range of 10^3–10^8.

Further, to enhance the sensitivity, a n+ pocket doped vertical tunnel field effect transistor label-free biosensor has been proposed [70]. This is also based on dielectric modulation technique for sensing applications. The sensitivity of the proposed vertical double gate TFET biosensor is 10^4 times more than the conventional-based double-gate TFET biosensor. The sensitivity achieved by this vertical TFET (VTFET) sensor is 1.56×10^6 and the double-gate conventional TFET is 1.05×10^2 for completely filled neutral biomolecules when K = 12. VTFET biosensor has a larger sensitivity due to the existence of lateral and vertical tunneling. Figure 4.14(a) and (b) [70] represents the schematic of conventional double-gate TFET biosensor and pocket doped double-gate vertical TFET biosensor, respectively. The addition of pocket doping in the source region generates vertical as well as horizontal tunneling and allows biomolecules to locate at various positions of the cavity and generates a noticeable I-V characteristic [71–73]. Figure 4.15(a) and (b) shows sensitivity plots with

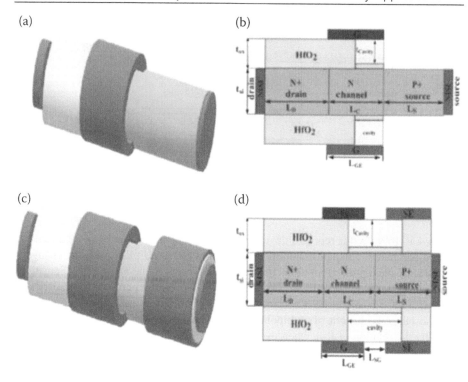

FIGURE 4.11 Schematic of devices: (a) 3D schematic of CNW TFET biosensor, (b) 2D schematic of CNW TFET biosensor, (c) 3D schematic of SE NW TFET biosensor, and (d) 2D schematic of SE NW TFET biosensor.

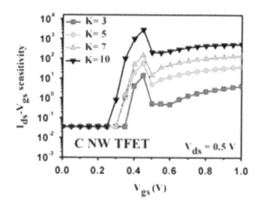

FIGURE 4.12 Current sensitivity plot for CNWTFET biosensor.

different k-values for conventional DG-TFET and pocket doped DG-vertical TFET biosensor [70]. Additionally, to understand the practicality of the sensor, investigation on steric hindrance is done. It is observed that decreasing partially filled profile followed by the concave step profile achieved maximum sensitivity among all the cases

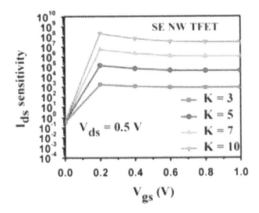

FIGURE 4.13 Current sensitivity plot for SE NWTFET biosensor.

[70]. The increasing partially filled profile followed by a convex profile has poor sensitivity among all. In this paper, [74] the author proposed a new sensing metric i.e., g_m/I_{ds} for TFET-based biosensor. This metric cannot only be used for sensitivity but also for selectivity for various biomolecules such as: streptavidin, biotin, and APTES whose K = 2.1, 2.63, and 3.57, respectively. This metric is taken into account of operation at low-power consumption (in the sub-threshold area), that is most prone to biomolecular detection. To improve the sensitivity of lateral gate metal work function engineered SiGe-source TFET biosensor, pocket doping is employed in the source area [75]. This author [75] uses subthreshold swing and drain current as a sensing parameter. To calculate the sensitivity through the subthreshold-swing it is given as

$$S_{SS} = \frac{SS_{air} - SS_{bio}}{SS_{air}} \qquad (4.4)$$

Here, SS_{air} and SS_{bio} are the subthreshold swing values in the existence of air and biomolecules in the etched dielectric portion, respectively. Further, to improve the random dopant fluctuations and to decrease the thermal budget charge, plasma-based TFET biosensors are designed.

4.5.2 Charge Plasma-Based Junctionless TFET Biosensors

Nanoscale devices are arising as a platform for sensing biomolecules. Different problems such as random doping variation and thermal budget have been identified throughout the manufacturing process of nanoscale devices. In order to remove these issues, the charge plasma (CP)-based concept is used in the TFET devices [5,76–80]. Blending the benefits of charge plasma with TFET-based dielectric modulated biosensor (label-free approach) resulted in the manufacture of a charge plasma-based

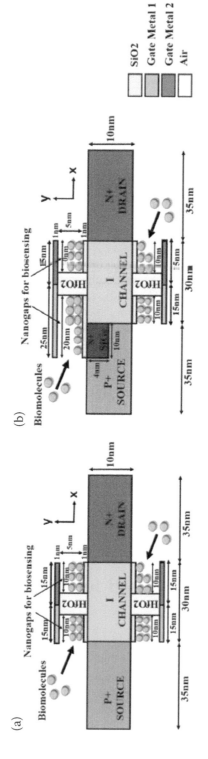

FIGURE 4.14 Schematic of (a) conventional double-gate TFET (b) pocket doped double-gate vertical TFET biosensor.

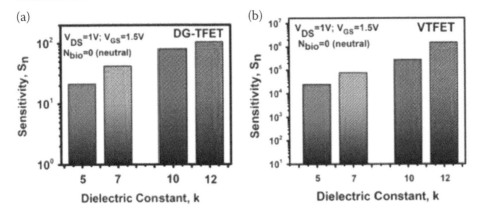

FIGURE 4.15 Sensitivity plots with different k-values (a) conventional DG-TFET (b) pocket doped DG-vertical TFET biosensor.

TFET biosensors [81]. Charge plasma tunnel FET is designed using suitable metal work-function for source and drain areas in intrinsic Si body (ni = 10^{15} cm^3). This technique eradicates the requirement of formation of junctions. Further, to increase the sensitivity, the author [82] reported a dual material charge plasma-based TFET biosensor and also discussed the influence of cavity length on the drain current sensitivity of the biosensor. The schematic view of conventional doping-less TFET and CP-based gate underlap DM- JLTFET are shown in Figure 4.16(a) and Figure 4.16(b), respectively [82]. The effect of cavity length variation on drain current chrs. (a) at V_{ds} = 0.5 V and (b) at V_{ds} = 1.5 V is shown in Figure 4.17 [82]. Similarly, [83] reported a label-free biosensor using a CP-based doping-less double-gate TFET device whose schematic is shown in Figure 4.18. The sensitivity obtained by this device is 10^{10}. Further, to increase the sensitivity of sensor research, various novel materials such as compound materials and 2D materials [84] etc. are used.

4.6 STRUCTURES AND WORKINGS OF TFET-BASED GAS SENSORS

H_2 gas sensors are constantly demanded in automotive, environmental, petroleum, and medical industries [85]. The basic operation behind the workings of a gas sensor is modulation in metal-gate work function when gases (such as H_2 and NH_3 etc.) react with the metal gate, which further causes variation in electrical characteristics such as V_{fb}, V_{th}, and I_d of the device [86]. This variation in conductivity is related to the quantity of an exact gas present in the environment. For detection of particular gas, a specific gate metal is used in the gas sensor because the sensitivity of a particular gate

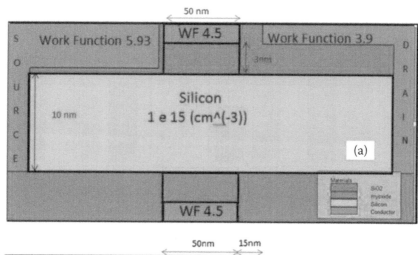

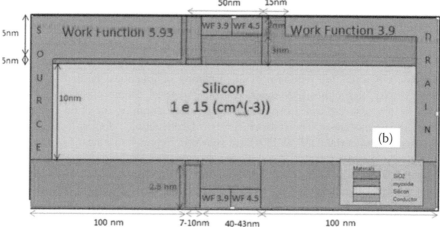

FIGURE 4.16 Cross-sectional view of (a) conventional doping-less TFET (b) charge plasma-based gate underlap DM- JLTFET.

metal material is affected by a particular gas like palladium gate electrode used for H_2 gas sensing and Ag is used for O_2 gas sensing [87,88].

A lot of work has been carried out with MOSFETs in the arena of semiconductor gas sensors. However, TFET must be examined as a gas sensor, as TFET is very well suited for low-power digital CMOS devices. TFET's unique features, such as low leakage current, decreased sub-threshold swing, and immunity from short channel effects make TFET a major candidate in the semiconductor industry for low-power devices. Besides these outstanding characteristics, the TFET suffers from several deficiencies, such as low I_{ON}, ambipolar current, and high threshold voltage. Variable engineering techniques, including hetero-gate dielectric techniques, gate metal work-function technique, heterojunction technique [59], gate drain underlap/overlap technique, etc., were applied to resolve these limitations of TFET. First, a TFET-based gas

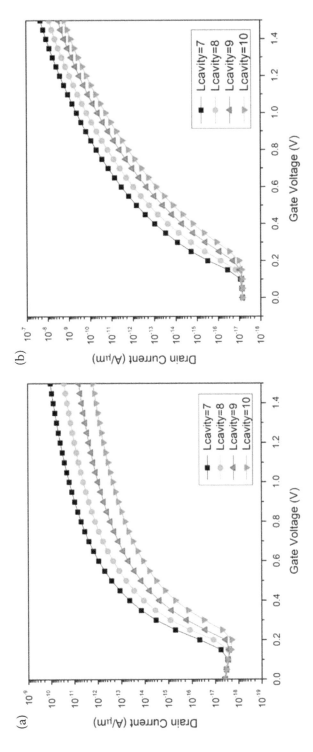

FIGURE 4.17 Effect of cavity length variation on drain current chrs. (a) at V_{ds} = 0.5 V and (b) at V_{ds} = 1.5 V.

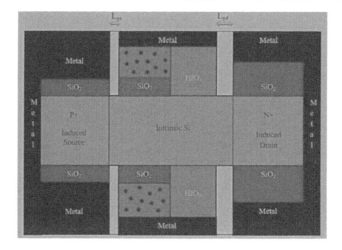

FIGURE 4.18 2D view of DLDGTFET biosensor.

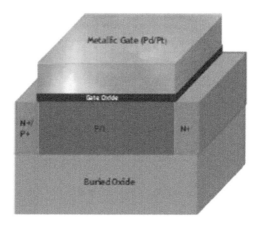

FIGURE 4.19 Schematic of metallic gate FET/TFET gas sensor.

sensor [89] which shows that TFET-based gas sensors have high sensitivity as compared to MOSFET-based gas sensor. The principle for TFET-based gas sensor is same as metallic gas sensor in which due to adsorption of gas molecules on the surface occurs and thus diffusion of gas molecules into metal gate, therefore some dipoles are formed and modulation of gate-metal work function takes place. Figure 4.19 shows the schematic of FET/TFET gas sensor based on a metallic gate [89]. Figure 4.20 shows the comparison plot for sensitivity of MOSFET and TFET-based gas sensor using a pladdium metal gate [89]. Sensitivity for gas sensors is the proportion of modulation of current after gas adsorption to the original current before gas adsorption. TFET-based gas sensors not only provide high sensitivity in comparision to conventional FET-based devices but also offer protection against screening against modulation of work-function due to non-specific gases.

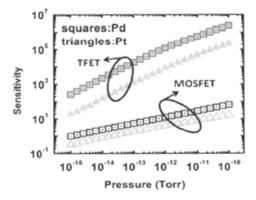

FIGURE 4.20 Comparative plot of sensitivity w.r.t pressure of MOSFET- and TFET-based gas sensors.

4.7 DIFFERENT ARCHITECTURES OF TFET-BASED GAS SENSORS

4.7.1 Metallic Gate Doped TFET-Based Gas Sensors

In this paper [90], the author proposed a tunnel FET-based gas sensor based on the palladium gate for H_2 gas sensing. The gate's all-around structure and hetero-dielectric scheme is employed for H_2 gas sensor to increase the sensitivity of the sensor. Figure 4.21 [90] shows (a) adsorption of gas molecules and (b) shows meshed schematic of a GAA hetero-dielectric gas sensor. The sensitivity of the gas sensor is calculated through equation. (4.5) [90].

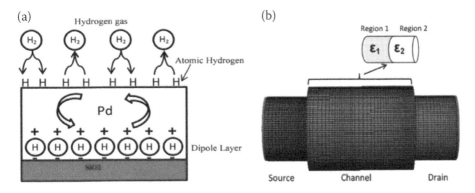

FIGURE 4.21 (a) Schematic of adsorption of gas molecules; (b) meshed schematic of HDGAA tunnel FET gas sensor.

$$S = \frac{I_{after\ gas\ adsorption} - I_{before\ gas\ adsorption}}{I_{before\ gas\ adsorption}} \qquad (4.5)$$

where $I_{after\ gas\ adsorption}$ and $I_{before\ the\ gas\ adsorption}$ is the value of current after adsorbing the gas molecules and before the adsorbtion of gas molecules, respectively. The sensitivity comparative plots for the conventional MOSFET, conventional TFET, GAA-TFET, and HD-GAA-TFET are shown in Figure 4.22. TFET sensitivity in either of the cases is greater than the traditional MOSFET throughout the full-pressure range, which is clearly analyzed from the result shown in Figure 4.22 [90]. The reason for increased sensitivity of GAATFET is that the cylindrical architecture's higher surface to volume relationship enhances the probability of H-atoms (dissociated) diffusion through the Pd and thus increases the quantity of dipoles created by the PdeSiO$_2$ interface to improve the sensitivity of the sensor.

4 7.2 Metallic Gate-Based Charge Plasma Based TFET Gas Sensors

Gas sensors particularly developed for human breath analysis are a reliable option for a fast and accurate detection of illnesses compared to existing conventional diagnosis procedures such as slow-through and intrusive blood tests. The shift in H_2S, NH_3, NO, and toluene trace species that are present in an exhaled breath may be utilized to detect halitosis, renal dysfunction, asthma, and lung cancer [15,91,92], correspondingly. Therefore, gas sensors need to be very sensitive for the purpose of an efficient biomedical diagnosis instrument. An ammonia gas sensing nanowire doping-less FET has

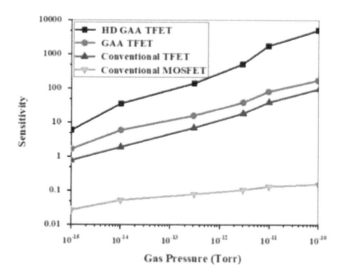

FIGURE 4.22 Sensitivity comparative plots of different cases of TFET and MOSFET.

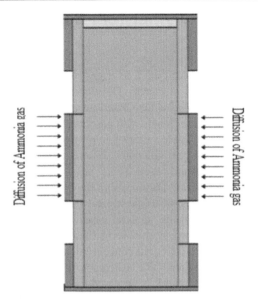

FIGURE 4.23 Schematic of IGZO VNWFET NH₃ gas sensor.

been reported in [93]. In this gas sensor, IGZO is used as a channel material, which has superior mobility as compared to conventional amorphous semiconductor material. The mechanism behind the gas sensing is that variation in work-function of gate metal (cobalt, molybdenum) as some gas molecules react with a metal gate that causes a variation in I_{off}, I_{on} and V_{th} parameters. These parameters are often taken into account as sensitivity parameters for NH₃ gas detection. The schematic for nanowire doping-less TFET is shown in Figure 4.23 [93]. A comparison plot of On-current sensitivity w.r.t modulation in work function for different catalytic gate metal electrodes is shown in Figure 4.24 [93]. Similarly, Sonal et.al. proposed a vertical junction-less TFET-based gas sensor for the detection of NH₃ gas [88].

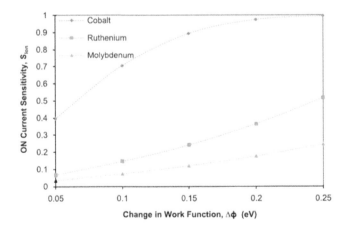

FIGURE 4.24 Comparision plot of on-current sensitivity w.r.t change in work function.

4.7.3 Conducting Polymer (CP)-Based GAA-TFET Gas Sensors

Research has been done using MOSFET as a gas sensor using catalytic metals and conducting polymers as a gate electrode. Developing a gas sensor using polymer conductor provides flexibility. This is also based on same principle i.e., modulation of gate work function as the polymer reacts with gas molecules. The advantage of using a conducting polymer is fast reversible variation in polymer work function that is further proportional to the quantity of gas molecules. Blending the advantage of a CP-based gas sensor with a gate all-around-based TFET provides a better power, extremely sensitive, and stable gas sensor that accomplishes the continuous demands of chemical and medicinal productions, biomedical investigation, environmental investigation, and automotive productions. Numerous analyte gases such as C_6H_{14}, CH_3OH, Iso-C_3H_7OH, CH_2Cl_2, and $CHCl_3$ have been observed through this sensor. The monitoring of such gases is essential as the increased concentration of such gases can damage the central nervous system of humans. The schematic of a CP-based GAA-TFET gas sensor is shown in Figure 4.25 [87]. It is observed that a PNIN GAA TFET based gas sensor has the highest sensitivity among all gas sensors [87].

4.8 STRUCTURES AND WORKINGS OF TFET-BASED PHOTOSENSOR

In the case of photosensors, most of the research is done on MOSFET devices until now [81,94–98]. The light has an impact on the potential drop of the tunneling gate, and there is an alteration in the channel's potential corresponding to the wavelength and amount of the light. This idea is applied to design an extremely sensitive photodetector. A cylindrical surrounding gate (CSG) MOSFET is proposed by [94] in which they used ZnO as a gate material and SiO_2 as dielectric. This device provides high

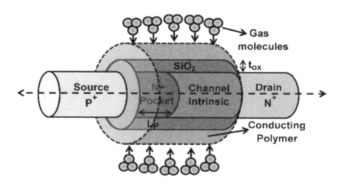

FIGURE 4.25 Schematic of PNIN-GAA-TFET-based gas sensor.

sensitivity over conventional double-gate MOSFET-based photosensors due to a cylindrical surrounding gate that provides high electrical coupling between the gate and channel, as the gate entirely surrounds the channel so leakage current is reduced and on-current (I_{on}) increases. The 2D and 3D view of a CSG photosensor is shown in Figure 4.26 and Figure 4.27 [94]. The author also investigates compound material engineering and work-function engineering on the photosensor device to enhance the sensitivity of the sensor [99].

A TFET device has the capability to reduce the short-channel effects and subthreshold swing, so researchers started considering the TFET device in photo-sensing applications because subthreshold swing and short channel effect can further affect the sensitivity of sensor, so subthreshold swing can further be used as a sensing metric for the device. The optical photosensor for infrared sensing using a TFET device [100] is shown in Figure 4.28. The workings behind this TFET-based photosensor device is

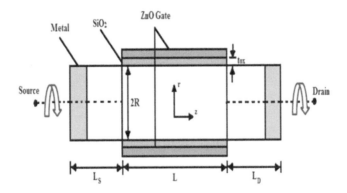

FIGURE 4.26 2D-view of CSG MOSFET.

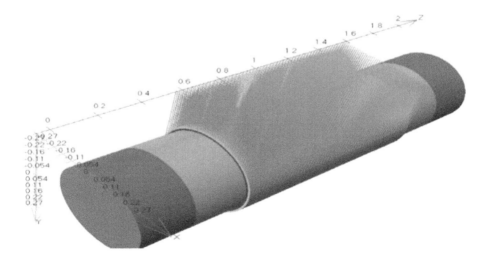

FIGURE 4.27 3-D view of CSG MOSFET under incident condition.

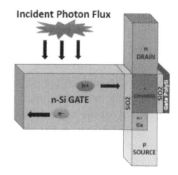

FIGURE 4.28 3-D view of optical photosensor device.

that during the illumination condition, when light incident normal to the illumination gate window electron hole pair generates. Photons have higher energy than the forbidden gap of Si that gets absorbed and this leads to photogeneration. When a positive gate bias is applied to the device, photogenerated e⁻ inside the gate gets attracted close to the electrode, and holes get collected close to the gate-dielectric edge. This procedure distracts the equilibrium in gate area and give rise to photo-voltage across it This photovoltage also acts as another gate bias that helps to lower the switch-on voltage needed for the BTBT process. Due to this, tunneling current during illumination increases. Further, a good amount of on-current can increase the speed as well as the sensitivity of the sensor. Furthermore, in this structure, Ge is used in the source region, which further impacts the I_{on} of the device that further increases the sensitivity of the sensor. Germanium is a possible option for the expansion of near-infrared silicon sensitivity from 1.3 μm to 1.55 μm and compatible with IC technology. The spectral sensitivity (Sn) [100] of the device, which is given in equation.(4.6), can be defined as the proportion of difference in drain current Id, when the illumination wavelength decreases from λb to λa, to the drain current of the original wavelength. This work does not take into account the non-ideal effects.

$$Sn = \frac{I_d(a)}{I_d(b)} - 1 \qquad (4.6)$$

4.9 DIFFERENT ARCHITECTURES OF TFET-BASED PHOTOSENSORS

4.9.1 Hybrid Photosensors

Photogating is a method to regulate the channel's conductance through light induced gate voltage [101] With the advancement in fabrication technology, nonstop

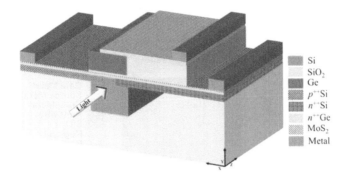

FIGURE 4.29 Schematic of a hybrid TFET photosensor.

progress in the development of TFET takes place by utilizing small-dimension materials like 2-D material transition metal dichalcogenide (TMD) materials, as TMDs have good scalability, small effective mass, and better sub-threshold characteristics. However, the optical absorption of TMDs is restricted by the thickness; the ultra-small thickness permits better gate adjustments. A hybrid TFET-based photosensor [102] is shown in Figure 4.29 in which germanium is used as an absorbing layer and MOS_2 is used as channel material. This is also based on the photogating principle. The Ge layer absorbs the light and photogate to MOS2 TFET. This photosensor design is a suitable candidate to detect the wide spectrum, even weak light with better performance. This device makes a balance between the optical as well as electrical properties that is required for the enhancement of the sensitivity of the photosensors. The I_{on}/I_{off} and responsivity at 1,500 nm wavelength (for infrared detection) obtained by this device is 10^7 and 2,700 AW^{-1}, respectively. Further, the author proposed a photosensor for visible light detection using split gate TMD TFET [35]. The obtained I_{on}/I_{Off} ratio by this split gate TMD TFET device is 10^8 and responsivity is in the range of $10^3 AmW^{-1}$ to 10^{-5} AmW^{-1} for $V_{gs} = 0.9$ V.

4.9.2 Hetero-Material-Based Charge Plasma-Based TFET Photosensors

Further, to solve the thermal budget and random-dopant fluctuations, the charge plasma-based TFET was proposed by [103]. In this photosensor device shown in Figure 4.30 [103], Cl-ITO (chlorinated-indium titanium oxide) used in region-I and source metal for enabling the illumination of photo-dielectric material is placed in region II. The variation in dielectric value occurs when brightened up by various light frequencies at static power. Improved current sensitivity to approximate 200% is achieved that provides a better choice to opt for photosensing applications.

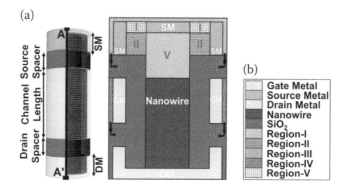

FIGURE 4.30 (a) 3-D view; (b) 2-D view of hetero-material-based charge plasma-based TFET device.

4.10 SUMMARY

Nano-sensor (biosensor, gas sensors, and photosensors)-based devices are required for the early detection and investigation methods to control the infection and cost of treatment of disease. Research and development of sensors is becoming the most widely studied area because the user-friendly, quick, low-cost, extremely sensitive, and extremely selective sensors contribute to advancement in next-generation medication i.e., customized medicine and highly sensitive point-of-care detection of disease markers. Sensitivity and response time are an important parameter for the biosensors. Highly sensitive biosensors are required for the accurate monitoring and treatment of the disease. From the above survey of biosensors, it is concluded that TFET-based biosensors are superior compared to traditional FET-based biosensors in terms of sensitivity and response time due to their different carrier injection mechanism i.e., band to band tunneling mechanism. Further, TFET-based sensors overcome the issue of short channel effects and power consumption too which were the limitation of MOSFET-based sensors. Further, improving the sensitivity of TFET-based sensors can be done by employing various engineerings in the device such as hetero-dielectric material engineering, short-gate technique, and pocket doping technique etc. Furthermore, to overcome the issue of fabrication, charge plasma technology is used for the TFET-based biosensors. In the recent time, diagnosis of COVID-19 pneumonia is done by chest CT scan test and RT-PCR test. Numerous suspected persons may not be verified and isolated in a timely manner due to overworked medical staff in epidemic areas, which is alarming. There is an urgent need for more reliable, quick-response, cost-effective, and widely accessible analytical equipment or diagnostic techniques. So, in this respect, biosensors are powerful tools to detect such viruses or bioterrorism threats. Gas sensors particularly developed for human breath analysis are a reliable option for a fast and accurate detection of illnesses compared to existing conventional diagnosis procedures such as slow and intrusive blood tests. Further monitoring of gases is also very essential, as over-concentration of few gases can damage the nervous system. Highly sensitive gas sensors can be designed using TFET devices and charge plasma technology. At the present time, the demand for low-power and extremely sensitive photosensors is rising at a surprising rate in numerous domains. For biomedical fields, highly sensitive photodetectors are required, so there we can use hybrid TMD-based TFET-based photosensors. This chapter provides researchers a quick review of previous reported biosensors, gas sensors, and photosensors until now and also gives an idea how to create future generation TFET-based different sensors with improved performance in terms of low-cost, highly sensitive, highly selective, more precise, more reliable, portable, and low response time.

REFERENCES

[1] N. Alizadeh, H. Jamalabadi, and F. Tavoli, "Breath Acetone Sensors as Non-Invasive Health Monitoring Systems: A Review," *IEEE Sensors Journal*, vol. 20, no. 1, pp. 5–31, 2020, doi: 10.1109/JSEN.2019.2942693

[2] "Optoelectronic sensors in medical applications," *components, fierce electronics*, 2003. https://www.fierceelectronics.com/components/optoelectronic-sensors-medical-applications

[3] S. P. Mohanty and E. Koucianos, "Biosensors: A tutorial review," *IEEE Potentials*, vol. 25, no. 2, pp. 35–40, 2006, doi: 10.1109/MP.2006.1649009

[4] P. Kumar, S. K. Sharma, and B. Raj, "Comparative Analysis of Nanowire Tunnel Field Effect Transistor for Biosensor Applications," *Silicon*, 2020, doi: 10.1007/s12633-020-00718-5

[5] G. Wadhwa and B. Raj, "Design, Simulation and Performance Analysis of JLTFET Biosensor for High Sensitivity," *Silicon*, vol. 18, pp. 567–574, 2019, doi: 10.1109/TNANO.2019.2918192

[6] T. Wadhera, D. Kakkar, and G. Wadhwa, "Recent Advances and Progress in Development of the Field Effect Transistor Biosensor: A Review," *Journal of Electronic Materials*, vol. 48, no. 12, pp. 7635–7646, 2019, doi: 10.1007/s11664-019-07705-6

[7] P. Bergveld, "Short Communications: Development of an Ion-Sensitive Solid-State Device for Neurophysiological Measurements," *IEEE Transactions on Biomedical Engineering*, vol. BME-17, no. 1, pp. 70–71, 1970, doi: 10.1109/TBME.1970.4502688

[8] A. Syahir, K. Usui, K. Tomizaki, K. Kajikawa, and H. Mihara, "Label and Label-Free Detection Techniques for Protein Microarrays," *Microarrays*, vol. 4, no. 2, pp. 228–244, 2015, doi: 10.3390/microarrays4020228

[9] A. Gao, N. Lu, Y. Wang, and T. Li, "Robust ultrasensitive tunneling-FET biosensor for point-of-care diagnostics," *Scientific Reports*, vol. 6, no. February, pp. 1–9, 2016, doi: 10.1038/srep22554

[10] B. G. Andryukov, N. N. Besednova, R. V. Romashko, T. S. Zaporozhets, and T. A. Efimov, "Label-free biosensors for laboratory-based diagnostics of infections: Current achievements and new trends," *Biosensors*, vol. 10, no. 2, pp. 1–22, 2020, doi: 10.3390/bios10020011

[11] N. N. Reddy and D. K. Panda, "A Comprehensive Review on Tunnel Field-Effect Transistor (TFET) Based Biosensors: Recent Advances and Future Prospects on Device Structure and Sensitivity," *Silicon 2020*, pp. 1–16, Aug. 2020, doi: 10.1007/S12633-020-00657-1

[12] A. Rhouati, G. Catanante, G. S. Nunes, and J. L. Marty, "Label-Free Aptasensors for the Detection of Mycotoxins," no. December, 2016, doi: 10.3390/s16122178

[13] S. Hong et al., "FET-type gas sensors: A review," Sensors Actuators, B Chem., vol. 330, no. October 2020, p. 129240, 2021, doi: 10.1016/j.snb.2020.129240

[14] S. Jamasb, "Continuous monitoring of pH and blood gases using ion-sensitive and gas-sensitive field effect transistors operating in the amperometric mode in presence of drift," Biosensors, vol. 9, no. 1, 2019, doi: 10.3390/bios9010044

[15] N. Nasiri and C. Clarke, "Nanostructured gas sensors for medical and health applications: Low to high dimensional materials," Biosensors, vol. 9, no. 1, pp. 1–22, 2019, doi: 10.3390/bios9010043

[16] A. G. Dent, T. G. Sutedja, and P. V. Zimmerman, "Exhaled breath analysis for lung cancer," Journal of Thoracic Disease, vol. 5, no. SUPPL.5, 2013, doi: 10.3978/j.issn.2 072-1439.2013.08.44

[17] M. Righettoni, A. Amann, and S. E. Pratsinis, "Breath analysis by nanostructured metal oxides as chemo-resistive gas sensors," Materials Today, vol. 18, no. 3, pp. 163–171, 2015, doi: 10.1016/j.mattod.2014.08.017

[18] M. Righettoni, A. Tricoli, and S. E. Pratsinis, "Si:WO3 sensors for highly selective detection of acetone for easy diagnosis of diabetes by breath analysis," AIChe Annual Meeting Conference Proceedings, vol. 82, no. 9, pp. 3581–3587, 2010.

[19] J. S. Jang, S. J. Choi, S. J. Kim, M. Hakim, and I. D. Kim, "Rational Design of Highly Porous SnO2 Nanotubes Functionalized with Biomimetic Nanocatalysts for Direct Observation of Simulated Diabetes," Advanced Functional Materials, vol. 26, no. 26, pp. 4740–4748, 2016, doi: 10.1002/adfm.201600797

[20] P. Fuchs, C. Loeseken, J. K. Schubert, and W. Miekisch, "Breath gas aldehydes as biomarkers of lung cancer," International Journal of Cancer, vol. 126, no. 11, pp. 2663–2670, Jun. 2010, doi: 10.1002/IJC.24970

[21] M. K. Nakhleh et al., "Diagnosis and Classification of 17 Diseases from 1404 Subjects via Pattern Analysis of Exhaled Molecules," ACS Nano, vol. 11, no. 1, pp. 112–125, Jan. 2016, doi: 10.1021/ACSNANO.6B04930

[22] S.-J. Choi et al., "Fast Responding Exhaled-Breath Sensors Using WO3 Hemitubes Functionalized by Graphene-Based Electronic Sensitizers for Diagnosis of Diseases," ACS Applied Materials & Interfaces, vol. 6, no. 12, pp. 9061–9070, Jun. 2014, doi: 10.1021/AM501394R

[23] X. A. Fu, M. Li, R. J. Knipp, M. H., Nantz, and M. Bousamra, "Noninvasive detection of lung cancer using exhaled breath," Cancer Medicine, vol. 3, no. 1, pp. 174–181, Feb. 2014, doi: 10.1002/CAM4.162

[24] B. Sharma and J.-S. Kim, "MEMS based highly sensitive dual FET gas sensor using graphene decorated Pd-Ag alloy nanoparticles for H2 detection," Scientific Reports 2018 81, vol. 8, no. 1, pp. 1–9, Apr. 2018, doi: 10.1038/s41598-018-24324-z

[25] E. Comini, G. Faglia, G. Sberveglieri, Z. Pan, and Z. L. Wang, "Stable and highly sensitive gas sensors based on semiconducting oxide nanobelts," Applied Physics Letters, vol. 81, no. 10, p. 1869, Aug. 2002, doi: 10.1063/1.1504867

[26] A. Paliwal, A. Sharma, M. Tomar, and V. Gupta, "Room temperature detection of NO2 gas using optical sensor based on surface plasmon resonance technique," Sensors Actuators B: Chemical., vol. 216, pp. 497–503, Sep. 2015, doi: 10.1016/J. SNB.2015. 03.095

[27] M. J. Tierney and H. O. L. Kim, "Electrochemical gas sensor with extremely fast response times," Analytical Chemistry, vol. 65, no. 23, pp. 3435–3440, Dec. 2002, doi: 10.1021/AC00071A017

[28] D. Sarkar et al., "A subthermionic tunnel field-effect transistor with an atomically thin channel," Nature, vol. 526, no. 7571, pp. 91–95, 2015, doi: 10.1038/nature15387

[29] A. Oprea, H. P. Frerichs, C. Wilbertz, M. Lehmann, and U. Weimar, "Hybrid gas sensor platform based on capacitive coupled field effect transistors: Ammonia and nitrogen dioxide detection," *Sensors Actuators B Chemical.*, vol. 127, no. 1, pp. 161–167, Oct. 2007, doi: 10.1016/J. SNB.2007.07.030

[30] J. Shin et al., "Thin-Wall Assembled SnO2 Fibers Functionalized by Catalytic Pt Nanoparticles and their Superior Exhaled-Breath-Sensing Properties for the Diagnosis of Diabetes," *Advanced Functional Materials*, vol. 23, no. 19, pp. 2357–2367, May 2013, doi: 10.1002/ADFM.201202729

[31] S. Bala and M. Khosla, "Design and analysis of electrostatic doped tunnel CNTFET for various process parameters variation," *Superlattices Microstructures.*, vol. 124, pp. 160–167, Dec. 2018, doi: 10.1016/J. SPMI.2018.10.007

[32] A. Singh, M. Khosla, and B. Raj, "Analysis of electrostatic doped Schottky barrier carbon nanotube FET for low power applications," *Journal of Materials Science: Materials in Electronics*, vol. 28, no. 2, pp. 1762–1768, Jan. 2017, doi: 10.1007/S10854-016-5723-7

[33] P. Zhang, Y. Xiao, J. Zhang, B. Liu, X. Ma, and Y. Wang, "Highly sensitive gas sensing platforms based on field effect Transistor-A review," *Analytica Chimica Acta*, vol. 1172, p. 338575, 2021, doi: 10.1016/j.aca.2021.338575

[34] O. JZ et al., "Physisorption-Based Charge Transfer in Two-Dimensional SnS2 for Selective and Reversible NO2 Gas Sensing," *ACS Nano*, vol. 9, no. 10, pp. 10313–10323, Oct. 2015, doi: 10.1021/ACSNANO.5B04343

[35] S. Joshi, P. K. Dubey, and B. K. Kaushik, "Photosensor Based on Split Gate TMD TFET Using Photogating Effect for Visible Light Detection," *IEEE Sensors Journal*, vol. 20, no. 12, pp. 6346–6353, 2020, doi: 10.1109/JSEN.2020.2966728

[36] R. Rubio et al., "Non-selective NDIR array for gas detection," *Sensors Actuators, B Chemical*, vol. 127, no. 1, pp. 69–73, Oct. 2007, doi: 10.1016/J. SNB.2007.07.003

[37] T. Hübert, L. Boon-Brett, G. Black, and U. Banach, "Hydrogen sensors – A review," *Sensors Actuators B Chemical*, vol. 157, no. 2, pp. 329–352, Oct. 2011, doi: 10.1016/J. SNB.2011.04.070

[38] U. Ackelid, M. Armgarth, A. Spetz, and I. Lundstroem, "Ethanol Sensitivity of Palladium-Gate Metal-Oxide-Semiconductor Structures.," *Electron device Letters*, vol. EDL-7, no. 6, pp. 353–355, 1986, doi: 10.1109/EDL.1986.26398

[39] F. Liao, C. Chen, and V. Subramanian, "Organic TFTs as gas sensors for electronic nose applications," *Sensors Actuators B Chemical*, vol. 107, no. 2, pp. 849–855, Jun. 2005, doi: 10.1016/J. SNB.2004.12.026

[40] A. Oprea et al., "Flip-chip suspended gate field effect transistors for ammonia detection," *Sensors Actuators B Chemical*, vol. 111–112, no. SUPPL., pp. 582–586, Nov. 2005, doi: 10.1016/J. SNB.2005.05.005

[41] J. H. Jeon and W. J. Cho, "Ultrasensitive Coplanar Dual-Gate ISFETs for Point-of-Care Biomedical Applications," *ACS Omega*, vol. 5, no. 22, pp. 12809–12815, 2020, doi: 10.1021/acsomega.0c00427

[42] C. H. Kim, I. T. Cho, J. M. Shin, K. B. Choi, J. K. Lee, and J. H. Lee, "A new gas sensor based on MOSFET having a horizontal floating-gate," *IEEE Electron Device Letters*, vol. 35, no. 2, pp. 265–267, Feb. 2014, doi: 10.1109/LED.2013.2294722

[43] C.-H. Lin and C. W. Liu, "Metal-Insulator-Semiconductor Photodetectors," *Sensors 2010, Vol. 10, Pages 8797-8826*, vol. 10, no. 10, pp. 8797–8826, Sep. 2010, doi: 10.3390/S101008797

[44] F. Gan, L. Hou, G. Wang, H. Liu, and J. Li, "Optical and recording properties of short wavelength optical storage materials," *Materials Science and Engineering B*, vol. 76, no. 1, pp. 63–68, Jun. 2000, doi: 10.1016/S0921-5107(00)00400-1

[45] K. Boucart and A. M. Ionescu, "A new definition of threshold voltage in Tunnel FETs," *Solid-State Electronics*, vol. 52, no. 9, pp. 1318–1323, 2008, doi: 10.1016/j.sse.2008.04.003

[46] A. C. Seabaugh and Q. Zhang, "Low-voltage tunnel transistors for beyond CMOS logic," *Proceedings of the IEEE*, vol. 98, no. 12, pp. 2095–2110, 2010, doi: 10.1109/JPROC.2010.2070470

[47] J. Knoch and J. Appenzeller, "A novel concept for field-effect transistors – The tunneling carbon nanotube FET," *Device Research Conference – Conference Digest, DRC*, vol. 2005. pp. 153–156, 2005, doi: 10.1109/DRC.2005.1553099

[48] S. Singh and B. Raj, "Study of parametric variations on hetero-junction vertical t-shape TFET for suppressing ambipolar conduction," *Indian Journal of Pure & Applied Physics*, vol. 58, no. 6, pp. 478–485, 2020.

[49] T. Chawla, M. Khosla, and B. Raj, "Design and simulation of triple metal double-gate germanium on insulator vertical tunnel field effect transistor," *Microelectronics Journal*, vol. 114, p. 105125, Aug. 2021, doi: 10.1016/J.MEJO.2021.105125

[50] E. Ko, H. Lee, J. D. Park, and C. Shin, "Vertical Tunnel FET: Design Optimization with Triple Metal-Gate Layers," *IEEE Transactions Electron Devices*, vol. 63, no. 12, pp. 5030–5035, 2016, doi: 10.1109/TED.2016.2619372

[51] S. Takagi and M. Takenaka, "III-V/Ge MOSFETs and TFETs for ultra-low power logic LSIs," *2017 Int. Symp. VLSI Technol. Syst. Appl. VLSI-TSA 2017*, pp. 3–4, 2017, doi: 10.1109/VLSI-TSA.2017.7942467

[52] W. Wang et al, "Design of U-shape channel tunnel FETs with SiGe source regions," *IEEE Transactions Electron Devices*, vol. 61, no. 1, pp. 193–197, 2014, doi: 10.1109/TED.2013.2289075

[53] S. Chen, S. Wang, H. Liu, W. Li, Q. Wang, and X. Wang, "Symmetric U-Shaped Gate Tunnel Field-Effect Transistor," *IEEE Transactions on Electron Devices* vol. 64, no. 3, pp. 1343–1349, 2017.

[54] S. W. Kim, J. H. Kim, T. J. K. Liu, W. Y. Choi, and B. G. Park, "Demonstration of L-Shaped Tunnel Field-Effect Transistors," *IEEE Transactions on Electron Devices* vol. 63, no. 4, pp. 1774–1778, 2015.

[55] Z. Yang, "Tunnel Field-Effect Transistor with an L-Shaped Gate," *IEEE Electron Device Letters.*, vol. 37, no. 7, pp. 839–842, 2016, doi: 10.1109/LED.2016.2574821

[56] H. Nam, M. H. Cho, and C. Shin, "Symmetric tunnel field-effect transistor (S-TFET)," *Current Applied Physics*, vol. 15, no. 2, pp. 71–77, 2015, doi: 10.1016/j.cap.2014.11.006

[57] H. Lee, S. Park, Y. Lee, H. Nam, and C. Shin, "Random variation analysis and variation-aware design of symmetric tunnel field-effect transistor," *IEEE Transactions Electron Devices*, vol. 62, no. 6, pp. 1778–1783, 2015, doi: 10.1109/TED.2014.2365805

[58] W. Li and J. C. S. Woo, "Optimization and Scaling of Ge-Pocket TFET," *IEEE Transactions Electron Devices*, vol. 65, no. 12, pp. 5289–5294, 2018, doi: 10.1109/TED.2018.2874047

[59] M. Mittal, M. Khosla, and T. Chawla, "Design and Performance Analysis of Delta-Doped Hetro-Dielectric GeOI Vertical TFET," *Silicon*, pp. 1–9, Aug. 2021, doi: 10.1007/S12633-021-01315-W

[60] P. K. Dubey and B. K. Kaushik, "T-Shaped III-V Heterojunction Tunneling Field-Effect Transistor," *IEEE Transactions Electron Devices*, vol. 64, no. 8, pp. 3120–3125, 2017, doi: 10.1109/TED.2017.2715853

[61] H. Lee, J. D. Park, and C. Shin, "Study of Random Variation in Germanium-Source Vertical Tunnel FET," *IEEE Transactions Electron Devices*, vol. 63, no. 5, pp. 1827–1834, 2016, doi: 10.1109/TED.2016.2539209

[62] W. Park, A. N. Hanna, A. T. Kutbee, and M. M. Hussain, "In-Line Tunnel Field Effect Transistor: Drive Current Improvement," *IEEE Journal of the Electron Devices Society*, vol. 6, no. March, pp. 721–725, 2018, doi: 10.1109/JEDS.2018.2844023

[63] M. Lee et al., "A novel biosensor based on hafnium oxide: Application for early stage detection of human interleukin-10," *Sensors Actuators B. Chemical*, vol. 175, pp. 201–207, 2012, doi: 10.1016/j.snb.2012.04.090

[64] D. Sarkar and K. Banerjee, "Proposal for tunnel-field-effect-transistor as ultra-sensitive and label-free biosensors," *Applied Physics Letters*, vol. 100, no. 14, 2012, doi: 10.1063/1.3698093

[65] D. Soni and D. Sharma, "Design of NW TFET biosensor for enhanced sensitivity and sensing speed by using cavity extension and additional source electrode," *Micro Nano Letters.*, vol. 14, no. 8, pp. 901–905, 2019, doi: 10.1049/mnl.2018.5733

[66] H. Im, X.-J. Huang, B. Gu, and Y.-K. Choi, "A dielectric-modulated field-effect transistor for biosensing," *Nature Nanotechnology 2007 27*, vol. 2, no. 7, pp. 430–434, Jun. 2007, doi: 10.1038/nnano.2007.180

[67] R. Narang, M. Saxena, R. S. Gupta, and M. Gupta, "Dielectric modulated tunnel field-effect transistor-a biomolecule sensor," *IEEE Electron Device Letters*, vol. 33, no. 2, pp. 266–268, 2012, doi: 10.1109/LED.2011.2174024

[68] S. Kanungo, S. Chattopadhyay, P. S. Gupta, and H. Rahaman, "Comparative performance analysis of the dielectrically modulated full- gate and short-gate tunnel FET-based biosensors," *IEEE Transactions Electron Devices*, vol. 62, no. 3, pp. 994–1001, 2015, doi: 10.1109/TED.2015.2390774

[69] D. B. Abdi and M. J. Kumar, "Dielectric modulated overlapping gate-on-drain tunnel-FET as a label-free biosensor," *Superlattices Microstructures*, vol. 86, pp. 198–202, 2015, doi: 10.1016/j.spmi.2015.07.052

[70] V. D. Wangkheirakpam, B. Bhowmick, and P. D. Pukhrambam, "N+ Pocket Doped Vertical TFET Based Dielectric-Modulated Biosensor Considering Non-Ideal Hybridization Issue: A Simulation Study," *IEEE Transactions on Nanotechnology*, vol. 19, pp. 156–162, 2020, doi: 10.1109/TNANO.2020.2969206

[71] P. Dwivedi and A. Kranti, "Overcoming Biomolecule Location-Dependent Sensitivity Degradation Through Point and Line Tunneling in Dielectric Modulated Biosensors," *IEEE Sensors Journal*, vol. 18, no. 23, pp. 9604–9611, Dec. 2018, doi: 10.1109/JSEN.2018.2872016

[72] P. Dwivedi, R. Singh, B. S. Sengar, A. Kumar, and V. Garg, "A New Simulation Approach of Transient Response to Enhance the Selectivity and Sensitivity in Tunneling Field Effect Transistor-Based Biosensor," *IEEE Sensors Journal*, vol. 21, no. 3, pp. 3201–3209, 2021, doi: 10.1109/JSEN.2020.3028153

[73] P. Dwivedi, R. Singh, and Y. S. Chauhan, "Crossing the Nernst Limit (59 mV/pH) of Sensitivity through Tunneling Transistor-Based Biosensor," *IEEE Sensors Journal*, vol. 21, no. 3, pp. 3233–3240, 2021, doi: 10.1109/JSEN.2020.3025975

[74] P. Dwivedi and A. Kranti, "Applicability of Transconductance-to-Current Ratio (gm/I_{ds}) as a Sensing Metric for Tunnel FET Biosensors," *IEEE Sensors Journal*, vol. 17, no. 4, pp. 1030–1036, 2017, doi: 10.1109/JSEN.2016.2640192

[75] M. Verma, S. Tirkey, S. Yadav, D. Sharma, and D. S. Yadav, "Performance Assessment of A Novel Vertical Dielectrically Modulated TFET-Based Biosensor," *IEEE Transactions Electron Devices*, vol. 64, no. 9, pp. 3841–3848, 2017, doi: 10.1109/TED.2017.2732820

[76] S. Singh, S. Bala, B. Raj, and B. Raj, "Improved Sensitivity of Dielectric Modulated Junctionless Transistor for Nanoscale Biosensor Design," *Sensor Letters*, vol. 18, no. 4, pp. 328–333, 2020, doi: 10.1166/sl.2020.4224

[77] S. Singh, B. Raj, and S. K. Vishvakarma, "Sensing and Bio-Sensing Research Analytical modeling of split-gate junction-less transistor for a biosensor application," *Sensing and Bio-Sensing Research*, vol. 18, no. February, pp. 31–36, 2018, doi: 10.1016/j.sbsr.2018.02.001

[78] S. K. Verma, S. Singh, G. Wadhwa, and B. Raj, "Detection of Biomolecules Using Charge – Plasma Based Gate Underlap Dielectric Modulated Dopingless TFET," *Transactions on Electrical and Electronic Materials*, no. 0123456789, 2020, doi: 10.1007/s42341-020-00205-z

[79] G. Wadhwa and B. Raj, "Surface Potential Modeling and Simulation Analysis of Dopingless TFET Biosensor," *Silicon 2021*, pp. 1–10, Feb. 2021, doi: 10.1007/S12 633-021-01011-9

[80] S. Alam, A. Raman, B. Raj, N. Kumar, and S. Singh, "Design and Analysis of Gate Overlapped/Underlapped NWFET Based Lable Free Biosensor," *Silicon*, 2021, doi: 10.1007/s12633-020-00880-w

[81] D. Singh, S. Pandey, K. Nigam, D. Sharma, D. S. Yadav, and P. Kondekar, "A charge-plasma-based dielectric-modulated junctionless TFET for biosensor label-free detection," *IEEE Transactions Electron Devices*, vol. 64, no. 1, pp. 271–278, 2017, doi: 10.1109/TED.2016.2622403

[82] G. Wadhwa and B. Raj, "Label Free Detection of Biomolecules Using Charge-Plasma-Based Gate Underlap Dielectric Modulated Junctionless TFET," *Journal of Electronic Materials*, vol. 47, no. 8, pp. 4683–4693, 2018, doi: 10.1007/s11664-018-6343-1

[83] S. Anand, A. Singh, S. I. Amin, and A. S. Thool, "Design and Performance Analysis of Dielectrically Modulated Doping-Less Tunnel FET-Based Label Free Biosensor," *IEEE Sensors Journal*, vol. 19, no. 12, pp. 4369–4374, 2019, doi: 10.1109/JSEN.201 9.2900092

[84] P. K. Dubey, N. Yogeswaran, F. Liu, A. Vilouras, B. K. Kaushik, and R. Dahiya, "Monolayer MoSe₂-Based tunneling field effect transistor for ultrasensitive strain sensing," *IEEE Transactions Electron Devices*, vol. 67, no. 5, pp. 2140–2146, 2020, doi: 10.1109/TED.2020.2982732

[85] N. K. Singh, A. Raman, S. Singh, and N. Kumar, "A novel high mobility In1-xGaxAs cylindrical-gate-nanowire FET for gas sensing application with enhanced sensitivity," *Superlattices Microstructures*, vol. 111, pp. 518–528, 2017, doi: 10.1016/j.spmi.201 7.07.001

[86] R. Gautam, M. Saxena, R. S. Gupta, and M. Gupta, "Gate-All-Around Nanowire MOSFET With Catalytic Metal Gate For Gas Sensing Applications," *IEEE Transactions on Nanotechnology*, vol. 12, no. 6, pp. 939–944, 2013, doi: 10.1109/ TNANO.2013.2276394

[87] J. Madan, R. Pandey, and R. Chaujar, "Conducting Polymer Based Gas Sensor Using PNIN- Gate All Around – Tunnel FET," *Silicon*, vol. 12, no. 12, pp. 2947–2955, 2020, doi: 10.1007/s12633-020-00394-5

[88] S. Singh, M. Khosla, G. Wadhwa, and B. Raj, "Design and analysis of double – gate junctionless vertical TFET for gas sensing applications," pp. 1–7, 2021.

[89] D. Sarkar, H. Gossner, W. Hansch, and K. Banerjee, "Tunnel-field-effect-transistor based gas-sensor: Introducing gas detection with a quantum-mechanical transducer," *Applied Physics Letters*, vol. 102, no. 2, 2013, doi: 10.1063/1.4775358

[90] J. Madan and R. Chaujar, "Palladium Gate All Around – Hetero Dielectric -Tunnel FET-based highly sensitive Hydrogen Gas Sensor," *Superlattices Microstructures*, vol. 100, pp. 401–408, 2016, doi: 10.1016/j.spmi.2016.09.050

[91] N. K. Singh, R. Kar, and D. Mandal, "Simulation and analysis of ZnO- based extended-gate gate-stack junctionless NWFET for hydrogen gas detection," *Applied Physics A: Materials Science and Processing*, vol. 127, no. 4, 2021, doi: 10.1007/s00339-021-04421-z

[92] S. Mokkapati, N. Jaiswal, M. Gupta, and A. Kranti, "Gate-All-Around Nanowire Junctionless Transistor-Based Hydrogen Gas Sensor," *IEEE Sensors Journal*, vol. 19, no. 13, pp. 4758–4764, 2019, doi: 10.1109/JSEN.2019.2903216

[93] N. Jayaswal, A. Raman, N. Kumar, and S. Singh, "Design and analysis of electrostatic-charge plasma based dopingless IGZO vertical nanowire FET for ammonia gas sensing," *Superlattices Microstructures*, vol. 125, no. October 2018, pp. 256–270, 2019, doi: 10.1016/j.spmi.2018.11.009

[94] A. Jain, S. K. Sharma, and B. Raj, "Design and analysis of high sensitivity photosensor using Cylindrical Surrounding Gate MOSFET for low power applications," *Engineering Science* and *Technology, an International Journal*, vol. 19, no. 4, pp. 1864–1870, Dec. 2016, doi: 10.1016/J.JESTCH.2016.08.013

[95] S. K. Sharma, B. Raj, and M. Khosla, "Enhanced photosensitivity of highly spectrum selective cylindrical gate In1−xGaxAs nanowire MOSFET photodetector," vol. 33, no. 12, May 2019, doi: 10.1142/S0217984919501446

[96] S. K. Sharma, P. Kumar, B. Raj, and B. Raj, "In1 − xGaxAs Double Metal Gate-Stacking Cylindrical Nanowire MOSFET for Highly Sensitive Photo Detector," *Silicon 2021*, pp. 1–7, May 2021, doi: 10.1007/S12633-021-01122-3

[97] S. K. Sharma, J. Singh, B. Raj, and M. Khosla, "Analysis of Barrier Layer Thickness on Performance of In 1− x Ga x As Based Gate Stack Cylindrical Gate Nanowire MOSFET," *Journal of Nanoelectronics and Optoelectronics*, vol. 13, no. 10, pp. 1473–1477, Nov. 2018, doi: 10.1166/JNO.2018.2374

[98] S. Kumar, B. Raj, and M. Khosla, "A Gaussian approach for analytical subthreshold current model of cylindrical nanowire FET with quantum mechanical effects," *Microelectronics Journal*, vol. 53, pp. 65–72, Jul. 2016, doi: 10.1016/J.MEJO.2016.04.002

[99] S. K. Sharma, A. Jain, and B. Raj, "Analysis of triple metal surrounding gate (TM-SG) III-V nanowire MOSFET for photosensing application," *Opto-Electronics Review*, vol. 26, no. 2, pp. 141–148, 2018, doi: 10.1016/j.opelre.2018.03.001

[100] V. Devi, W. Brinda, B. Puspa, and D. Pukhrambam, "Near − infrared optical sensor based on band − to − band tunnel FET," pp. 1–9, 2019.

[101] H. Fang and W. Hu, "Photogating in Low Dimensional Photodetectors," *Advanced Science*, vol. 4, no. 12, Dec. 2017, doi: 10.1002/ADVS.201700323

[102] S. Joshi, P. K. Dubey, and B. K. Kaushik, "A Transition Metal Dichalcogenide Tunnel FET-Based Waveguide-Integrated Photodetector Using Ge for Near-Infrared Detection," *IEEE Sensors Journal*, vol. 19, no. 20, pp. 9187–9193, 2019, doi: 10.1109/JSEN.2019.2922250

[103] N. Kumar and A. Raman, "Prospective Sensing Applications of Novel Heteromaterial Based Dopingless Nanowire-TFET at Low Operating Voltage," *IEEE Transactions on Nanotechnology*, vol. 19, pp. 527–534, 2020, doi: 10.1109/TNANO.2020.3005026

Modeling and Simulation Analysis of TFET-Based Devices for Biosensor Applications

5

M. Arun Kumar and Meenakshi Devi

Contents

5.1	Introduction to Tunneling Field Effect Transistor (TFET) Devices	106
5.2	Types of TFET Devices	107
	5.2.1 Dielectric Modulated Double Source Trench Gate Tunnel FET (DM-DSTGTFET)	107
	5.2.1.1 DM-DSTGTFET Device Structure	107
	5.2.1.2 Simulation Method and Model	108
	5.2.1.3 Device Performance	108
	5.2.2 DM-GUD-TFET (Dielectric Modulated Gate Underlap Doping-Less Tunnel Field Effect Transistor)	110
	5.2.2.1 DM-GUD-TFET Device Structure	110
	5.2.2.2 Device Performance	111
	5.2.3 Dielectric Modulated p-Type Tunnel Field Effect Transistor (DM p-TFET)	113
	5.2.3.1 Device Structure and Simulation Setup	113

DOI: 10.1201/9781003189633-5

	5.2.3.2 Device Performance	114
5.2.4	DMFET (Dielectrically Modulated Field Effect Transistor)	115
	5.2.4.1 Architecture and Simulation of TFET Biosensors	115
	5.2.4.2 Sensitivity Analysis of the TFET Device	117
5.3 Conclusion		118
References		118

5.1 INTRODUCTION TO TUNNELING FIELD EFFECT TRANSISTOR (TFET) DEVICES

Many research activities have concentrated on silicon-based field effect transistor biosensors to achieve further high sensitivity, scaled dimensions, low cost, minimum delay, and low-power consumption in biomedical applications [1,2]. The thermal electron emission is one of the limitations of FET-based biosensors with subthreshold slopes of not more than 60 mV/decade. The tunneling field effect transistor (TFET) overcomes the above limitation due to the band to band tunneling (BTBT) conduction mechanism [3]. Therefore, the TFET biosensor is preferable in biomedical applications because it has high sensitivity and quicker response than normal FET biosensors [4,5]. The detection of biomolecules in the TFET biosensor is dependent upon dielectric modulation. The two cavities are created in the gate dielectric material to receive the biomolecules. The dielectric constant of the cavities is changed when it is filled with biomolecules, resulting in changes in the transfer characteristics and drain current [6].

The dielectric modulated tunneling field effect transistor (DMTFET) biosensors have more value because they sense neutral and charged molecules [7]. The double channel trench gate field effect transistor-based biosensor provides high current and voltage sensitivity [8]. The sensitivity can also vary by implementing double metal materials and double-gate TFET biosensors [9]. The dielectric modulated dual-source trench gate tunneling field effect transistor (DM-DSTGTFET) [10] biosensor has higher sensitivity than other sensors because it has a large gate-source overlapping area. In this DM-DSTGTFET biosensor, the biomolecules are loaded vertically into the cavities from the top end. DM-DSTGTFET biosensor embraces a trench gate and dual source to generate bidirectional current and enrich the on-state current. It has two cavities etched more than 1 nm to fill the biomolecules. The capability of the biosensor (DM-DSTGTFET) consumes low power and provides high sensitivity. This property of the biosensor is highly suitable for biomedical applications.

In some cases, to increase the device's sensitivity, the device's structure was modified, introducing the DM-GUD-TFET (dielectric modulated gate underlap doping-less tunnel field effect transistor) [11] biosensor. TFET is a semiconductor device that is like MOSFET but uses quantum tunneling for conduction.

5.2 TYPES OF TFET DEVICES

5.2.1 Dielectric Modulated Double Source Trench Gate Tunnel FET (DM-DSTGTFET)

Biosensors based on a dielectric modulated double source trench gate tunnel FET (DM-DSTGTFET) are used to detect the biomolecules. A biosensor embraces a trench gate and dual source to generate bidirectional current and enrich the on-state current. It has two cavities etched more than 1 nm to fill the biomolecules. The capability of the biosensor (DM-DSTGTFET) consumes low power and provides high sensitivity. The property of this biosensor is highly suitable for biomedical applications. The device structure and the performances are discussed in the following section.

5.2.1.1 DM-DSTGTFET Device Structure

The schematic cross-sectional view of the DM-DSTGTFET biosensor is shown in Figure 5.1. The dual-source structure of TFET increases the on-state current. The two cavities are located symmetrically on the two sides of the gate material with a doping concentration of 1×10^{20} cm^{-3}. The p-channel is located below the gate material with a height of 37 nm and having a doping concentration of 1×10^{15} cm^{-3}. The n-channel is situated below the p-channel with a size of 18 nm and has a doping concentration of 1×10^{17} cm^{-3}. The pocket regions are placed symmetrically on the two sides of the gate with a thickness of 5 nm and has a donor doping concentration of 1×10^{19} cm^{-3}. The two oxides are HfO$_2$ is placed on the source region having a thickness of 2 nm. The metalwork function above the HfO$_2$ gate oxide is around Φ_{MS} = 4.2 eV. The proposed biosensor's electrical characteristics will analyze five different thicknesses of cavities (5 nm, 7 nm, 9 nm, 11 nm, 13 nm) and biomolecules with five different dielectric constants (1, 2.5, 5, 11, 23).

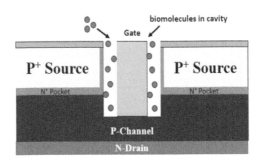

FIGURE 5.1 Shows the schematic cross-sectional view of DM-DSTGTFET.

5.2.1.2 Simulation Method and Model

The performance of the DM-DSTGTFET biosensor structure, the TCAD tool (sentaurus), is used to examine the sensitivity of the TFET-based sensor devices. The electric field measures the simulation results of the devices at each tunneling junction. The tunneling BTBT models will measure the actual situation of the TFET simulation. Similarly, the Kane model is also used to measure the BTBT tunneling in the sentaurus tool. At the same time, the device measurement notifies some essential factors. Shockley-Read-Hall (SRH) is to be considered to include the recombination of carriers. The bandgap narrowing model is also to be considered for the concentration effect in the bandgap.

Furthermore, Fermi-Dirac statics is used to measure the change in properties. To analyze the sensitivity of the DM-DSTGTFET biosensor device, parameters such as voltage swing, subthreshold swing, thickness, threshold voltage sensitivity, a cavity filled with air and biomolecules should be considered. However, few of the parameters were discussed in the device performance section.

5.2.1.3 Device Performance

The proposed DM-DSTGTFET biosensor's performance analysis will be studied under three different categories based on varying the dielectric constant of the cavities, varying the thickness of the cavities, and charged biomolecules.

1. *Impact of different dielectric constant of the cavities on DM-DSTGTFET*

The device performance will show the on-state characteristics like energy band variations, current sensitivity, transfer characteristics, and threshold voltage sensitivity with biomolecules at different dielectric constants. The metal gate work function is fixed to a lower value, $\Phi MS = 4.2$ eV, to study the features. The characteristics of the drain current with a different dielectric constant are shown in Figure 5.2. In this type of

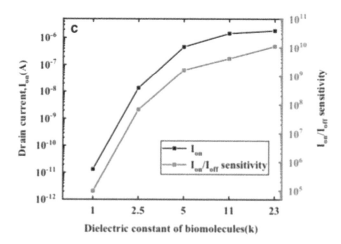

FIGURE 5.2 Current sensitivity of the device.

device, the drain current also increases when the gate voltage increases [11]. The energy band will be varied for the different dielectric constants. Whenever no biomolecules are present in the cavities (k = 1), the energy band has a minimal twist. The dielectric constant of the cavities is increased when it is filled with biomolecules.

At a higher dielectric constant, the energy band bends more severely. Therefore, the width of the barrier across the junction is increased because of more alignment in the energy band. Moreover, the current sensitivity of these devices will be achieved I_{on} and I_{on}/I_{off} by the dielectric constant. While increasing the dielectric constant, the current sensitivity increases because the barrier's width across the source-channel junction is decreased. The highest current sensitivity of 1.1×10^{10} is achieved at dielectric constant k = 23. From these results, it is to be understood that the impact of the dielectric constant of the cavities improves the performances of the devices.

2. Impact of different thicknesses of the cavities on DM-DSTGTFET

The impact over the thickness of the cavities is studied at a fixed dielectric constant value k = 23. The transfer characteristics, current sensitivity, and threshold voltage sensitivity of the proposed biosensor are discussed with various thicknesses of the cavities. The transfer characteristics of the DM-DSTGTFET biosensor with different thickness of the cavities is shown in Figure 5.3. The drain current will reduce when the thickness of the cavity is increased. Moreover, the on-state current is reduced when the cavities' thickness increases, which results in increased threshold sensitivity values. The gradual increase of cavity thickness reduces the control of the gate over the channel. Hence the performance of the device also depends upon the thickness of the cavities.

3. Impact of charged biomolecules on DM-DSTGTFET

The impact of positive and negatively charged biomolecules on DM-DSTGTFET biosensors is discussed in this section. The sensitivity of the charge density of this biosensor device ranges from 10^{10} cm^{-3} to 10^{13} cm^{-3}. DM-DSTGTFET-based biosensor

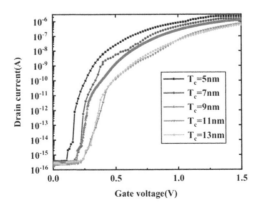

FIGURE 5.3 The transfer characteristics of the biosensor with various cavity thicknesses.

has a broader range of charge densities than the other sensors [12]. Therefore, the charge density within the dynamic limit range is used for sensitivity research. The positive charge of biomolecules was increased, which led to improving the device's Ion and Ion/Ioff sensitivity.

Furthermore, the positively charged biomolecules in the rise in cavity the gate oxide effect in the dielectric, which helps to improve the gate control ability. On the other hand, if the negative charge of biomolecules were increased, it would decrease the device's sensitivity and reduce the gate control ability of the devices. Hence, the DM-DSTGTFET has high sensitivity, which is used to detect the biomolecules in biosensor applications. The DM-DSTGTFET device structure was examined by the effects of relative permittivity, cavity thickness, charged biomolecules, and sensitivity. The discussion clearly shows that the DM-DSTGTFET device is applicable for the ultra-sensitive biosensor device.

5.2.2 DM-GUD-TFET (Dielectric Modulated Gate Underlap Doping-Less Tunnel Field Effect Transistor)

Semiconductor devices are tremendously used to build biosensors as they are abundantly available, susceptible, reliable, fast, and trim. ISFET (ion sensitive field effect transistor) is a typical semiconductor biotransducer. Still, it has few limitations, such as long-term hysteresis, external noise, intrinsic noise, and low recognition ability towards neutral analytes. To overcome these problems, DMFET (dielectric modulated field effect transistor) was introduced. DMFET can detect charged and uncharged analytes, but short channel effects are still present due to miniaturization [13]. To reduce SCEs and increase sensitivity, DM-GUD-TFET (dielectric modulated gate underlap doping less tunnel field effect transistor) biosensor is proposed. TFET is a semiconductor device that is like MOSFET but uses quantum tunneling for conduction. DM-GUD-TFET biosensor is a label-free sensor. Label-free sensors have high specificity and sensitivity; only one reagent is sufficient for the analysis; therefore, the cost and analysis time are reduced compared to the labeled methodology [14]. Immobilization of the target biomolecules is carried out to stabilize the biomolecule such that it is coupled to the bioreceptor, and the sensing process is initiated. Various immobilization methods are adsorption, covalent binding, cross-linking, and entrapment. Adsorption is most widely used for the label-free method.

5.2.2.1 DM-GUD-TFET Device Structure

The device structure of the DM-GUD-TFET is shown in Figure 5.4. In this structure, a nanocavity is created on the side of the gate metal to get high sensitivity for biomedical applications. The surface potential of the FET is changed when the biomolecules are immobilized in the cavity. The device variables for DM-GUD-TFET are channel length L_{ch} (50 nm), substrate thickness T_{su} (10 nm), channel doping ($1 \times 10^{10} cm^{-3}$), spacer thickness L_{us} and L_{ud} (3 nm and 15 nm), the work function of tunneling gate

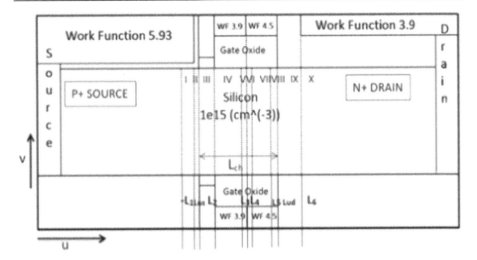

FIGURE 5.4 Schematic diagram of the DM-GUD-TFET.

metal, secondary gate metal are (3.9 eV and 4.5 eV), and length of the cavity (7 nm). The surface potential of the device was analyzed by dividing the structure into 10 sections, including source-drain prospects. This one-dimensional Poisson's equation is investigated for the underlap area and source-drain depleted section. The 10 sections of the device are denoted by I,II,III,IV,V,VI,VII,VIII,IX,X, as shown in the device structure. Here, I and X sections are the source and drain depletion sections, II and IX are gate underlap sections between gate and source metal, V and VI are depletion regions at the intersection between two gate metals, III and VIII are drift-diffusion portions, and IV and VII are gate section. The solution for the double-gate doping-less TFET is given by the one-dimensional Poisson equation. The potential of section VIII is analyzed using Poisson's 2-D equation. The solution is

$$\Psi n\,(u) = A_I \exp\frac{u - u_n}{D_I} + B_I \exp\left(-\frac{(u - u_n)}{D_I}\right) + \Psi gn\frac{-q.\,Nn.\,Tch}{C_{ox}} \quad (5.1)$$

By considering the boundary equations for equation (5.1), the surface potentials of all the sections are analyzed.

5.2.2.2 Device Performance

The performance of the device was analyzed by the simulation results SILVACO 2D ATLAS platform. In the simulation results, the variation of surface potential is plotted for neutral and charged biomolecules ($N_f = -9e15$ to $9e15$) at K = 5, V_{ds} = 0.5 v, Vgd = 0.35 v. When a negatively charged biomolecule is present in the cavity, the width of the depletion region increases at the source channel intersection, flat band voltage increases, and actual gate bias is decreased; therefore, the effective potential of the channel is decreased. When a positively charged biomolecule is present in the cavity width of the depletion region decreases at the source channel intersection, flat band voltage decreases, and actual gate bias increases (as shown in Figures 5.5 and 5.6). Therefore, the effective potential of the channel is increased. The potential increases when the neutral and positive biomolecule is introduced in the nanocavity, as shown below [15–17]. If the spacer length L_{us} at the gate side increases, the tunneling level decreases and affects the potential. If the spacer length is increased from 2 nm to 5 nm, the source channel intersection is reduced, and barrier width increases, reducing the tunneling effect. So, spacer length should be small for better characteristics [18].

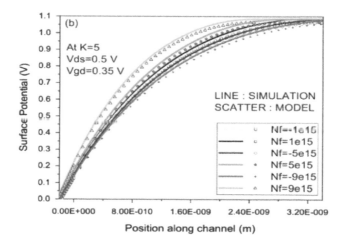

FIGURE 5.5 Simulated and modeled potential outcomes of the DM-GUD-TFET with charged biomolecules.

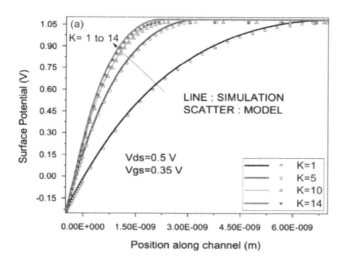

FIGURE 5.6 Simulated and modeled potential outcomes of the DM-GUD-TFET with neutral biomolecules.

Hence, the DM-GUD-TFET device structure has high performance with a modest choice of spacer region widths and thickness of tunneling gate length with proper biasing conditions. The DM-GUD-TFET has fewer short channel effects and leakage currents. This structure is less expensive and efficient in the development of biosensors. Thus, the device is susceptible, so it is ultrasensitive and cost-effective biomedical equipment.

5.2.3 Dielectric Modulated p-Type Tunnel Field Effect Transistor (DM p-TFET)

TFET is considered a promising alternative to MOSFET for dielectrically modulated biosensing applications. Tunnel FET is the best alternative for supply voltage reduction while maintaining a steep sub-threshold slope [19,20]. The DM TFET shows the superior electrical response over DM MOSFET due to its band to band tunneling and carrier injection mechanism compared to the thermionic injection mechanism of its DM MOSFET. The DM TFET usually incorporates a nanogap cavity in the gate metal or gate insulator region where the appropriate bioreceptors are activated. Subsequently, the target biomolecules (analyte) are introduced that conjugate with receptors and change either the dielectric constant or the charge density or both within the cavity region [21]. This alters the gating effect that leads to a modulation in the electrical characteristics of DM TFET, depending on the properties of the analyte/receptor binding system. Hence, DM FET-based biosensors detect charged as well as neutral biomolecules. Sensitivity and selectivity are the two essential sensing metrics for FET biosensors [22]. Herein, biomolecules such as APTES, biotin, and streptavidin through TFET with enhanced sensing metrics were discussed. Here, the tunability of sensing metrics through the optimization of gate back bias and gate work function is also addressed.

5.2.3.1 Device Structure and Simulation Setup

A single cavity DM P-TFET biosensor is shown in Figure 5.7. The device structure consists of a buried oxide (T_{box}) of 20 nm, silicon film thickness (T_{si}) of 10 nm, and oxide thickness (T_{ox}) of 10 nm. The gate length (L_g) of 200 nm with a nanogap cavity length of 75 nm towards sourced channel junction is considered. The biomolecules' APTES, biotin, and streptavidin has a thickness of 0.9 nm, 0.6 nm, and 6.1 nm, respectively [23], so a cavity of thickness 10 nm is chosen to detect any protein that lies

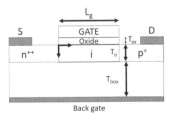

FIGURE 5.7 Schematic diagram of the DM p-TFET.

within the 10 nm range. The electric constant (ϵ_k) of biomolecules' APTES, biotin, and streptavidin are 3.57, 2.63, and 2.1, respectively. The variation in electrical characteristics due to neutral biomolecules can be studied by uniformly filling the cavity with a biomolecule corresponding to a specified dielectric permittivity. To obtain the device performance, bandgap narrowing and SRH processes [24] are considered. The Kane model is adopted to analyze band-to-band tunneling in p-TFET. The dry environment is preferred as the biosensor has sensitivity limitations under an aqueous environment due to the high ion concentration of solutions [25].

5.2.3.2 Device Performance

For every TFET device, functioning as absolute sensor I_{ds} is not important, but enhancing the difference in drain current in the presence and absence of biomolecule is required. The two important sensing metrics of the FET biosensor are sensitivity and selectivity. Sensitivity (S) is defined as the ratio of drain current ($I_{ds,bio}/I_{ds,air}$) in the presence and absence of biomolecules [26]. Figure 5.8(a) compares sensitivity with gate front voltage for different values of gate work function ($\emptyset_m$). It can be inferred that sensitivity is maximum at lower work function of $\emptyset_m = 4.2$ eV while the minimum at higher work function values of 5 eV. Peak sensitivity at $\emptyset_m = 4.2$ eV is nearly × 10 and × 50 times larger than that obtained at $\emptyset_m = 4.6$ eV and 5 eV, respectively. Figure 5.8(b) shows the dependency of S_{max} on dielectric permittivity (ϵ_k) of biomolecules for given values of $\emptyset_m$. A DM TFET-based sensor exhibited higher sensitivity at lower gate function, indicating the increase in the drain current in the presence of biomolecules in the cavity. Hence, to enhance the maximum sensitivity of the TFET biosensor work function, optimization is important.

The impact of backgate bias (V_{bg}) on the maximum drain current sensitivity (S_{max}) is required. The device's electrical characteristics can explain the dependency of maximum sensitivity on backgate bias. For example, the negative back bias on p-TFET reduces barrier width between source and drain and results in tunneling many carriers

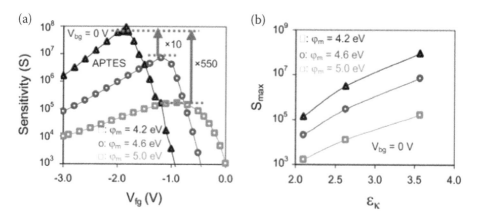

FIGURE 5.8 (a) Comparison of sensitivity (S) of APTES with V_{fg}; (b) maximum sensitivity on dielectric permittivity.

from source to channel. This increases the drain current in the absence of biomolecules and reduces the sensitivity as the relative difference in I_{ds} is diminished due to the presence of biomolecules in the cavity. In contrast, if a positive back bias up to 1 V decreases I_{ds} due to wider tunneling width. Therefore, the relative change in I_{ds} due to the presence of biomolecules is maximized, and sensitivity improves.

The dielectric constant of biomolecules from low- to high-value I_{ds} increases due to increased gate controllability over the channel [19]. The sensitivity also increases with the rise of $€_k$ of biomolecules because the ratio between $I_{ds,bio}/I_{ds, the\ air}$ changed to a higher value. Another important metric for TFET biosensors is selectivity (ΔS), distinguishing various biomolecules [27]. In practical applications of the FET-based biosensor, the location of biomolecules is essential as they are disordered in the partially filled cavity and may affect the functioning of the sensor. To analyze the effect, junction tunneling will be considered.

5.2.4 DMFET (Dielectrically Modulated Field Effect Transistor)

The concept of DMFET (dielectrically modulated field effect transistor) has been established as a constructive way to solve the critical concerns about the detection process [28]. Also, the inception of the biomolecule to the nanogap cavity made within the gate oxide or metal electrode will respond to the change in the electrostatics [29]. Corresponding to the specific biomolecule, the dielectric constant will change from k = 1 to k_{bio} in the DM concept at the entrance. The biomolecule may get a negative or neutral charge during conjugation. And in any case, the channel conduction will change concerning the channel electrostatics. The key parameter to estimate biosensor sensitivity may be variations in current, transconductance, or threshold voltage. To dominate the challenge of reduced sensitivity while conjugating the biomolecules, the TFET uses the BTBT mechanism, proven to be the practical solution [30]. The TAT component causes low sensing effects on the sensing of TFET-based biosensors.

Two different types of TFET biosensors have been proposed recently, where first includes a SiGe source, and the other has a doped pocket in its channel. Any of them may increase the BTBT rate, which increases the sensitivity. In any of the cases, the drawback is its low current during sensing performance. SiGe source will reduce bandgap and tunneling length, whereas doped pocket will decrease only the tunneling length.

5.2.4.1 Architecture and Simulation of TFET Biosensors

The cross-sectional view of the TFET biosensor is shown in Figure 5.9. Here, the double-gate configuration is applied to get the high control on electrostatics of the channel. For biomolecule trapping, NC is placed inside the gate oxide. When the fluid containing biomolecule enters the NC, it is joined to the related receptor, increasing the DC to K_{bio}>1. Thus, the channel conduction increases as the potential increases. Hence, the biomolecule is detected using the deflection in the current.

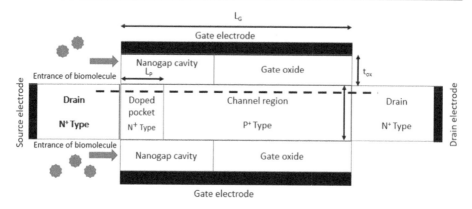

FIGURE 5.9 Schematic diagram of the TFET structure.

For the sensitivity calculations, $K_{bio} = 2.1$ is assumed here. The particular molecule corresponds to the biotin-streptavidin binding system. This binding is generally used in the literature of biomolecules. Interaction between streptavidin and biotin is one of the most vital non-covalent interactions [31]. To reduce the bandgap, the source region is made up of SiGe. Also, the n+ doped pocket is of length 10 nm. Therefore, SiGe material and doped pocket introduction will lead to a decrease in the bandgap and tunneling length. The conclusion can be made as an increase in BTBT thus increases BS. As the SiGe material and doped pocket are applied here, the structure may be called the novel structure as SIG-DP TFET biosensor. This structure is also compared with two other structures: one has only doped pockets, and the other does not have a doped pocket or SiGe material. Table 5.1 shows a list of parameters used to design the SIG-DP FET biosensor to detect the biomolecule in the simulation tool.

TABLE 5.1 List of the SIG-DP TFET biosensor parameters implemented in the simulation domain

PARAMETERS	VALUES
Gate length, L_g	100 nm
Source/drain region length, (L_s, L_d)	50 nm
Gate oxide thickness, t_{ox}	10 nm
Nanogap cavity length, L_{gap}	15 nm
Thickness of nanogap	10 nm
Silicon layer thickness, t_{si}	20 nm
Doping concentration of channel, (P⁻ type)	$1 \times 10^{16} cm^{-3}$
Doping concentration of source, (P⁺⁺ type)	$1 \times 10^{20} cm^{-3}$
Doping concentration of drain, (N⁺ type)	$5 \times 10^{18} cm^{-3}$
Gate metal workfunction	4.1 eV
Mole fraction Ge	0.3
Length of doped pocket, L_p	10 nm
Doping concentration of doped pocket, (N⁺ type)	$5 \times 10^{18} cm^{-3}$

For modeling and simulation, the SILVACO [32] family was used to design the structure. Necessary materials like SiGe and Si are applied in a simulator to get the output. BTBT nonlocal mechanism is considered here to get the realistic performance of the biosensor as the electrical performance of TFET is ruled by the tunneling component. The TAT component is also activated in the simulator as the TFET biosensor also has process-dependent issues. The other parameters such as BGN, FLDMOB, drift diffusion model, and Fermi statics are considered during the biosensor simulation [32].

Figure 5.10 shows the steps for the fabrication of the biosensor [33–63]. Here, two wafers, A and B, will follow the same flow steps. The starting wafer in each case is 5 nm thick,and a doping concentration of 1 e-16 cm-3 is creatoed. Nanogap creation is the most critical step. BTBT is represented as a function of Ge and doped pocket concentration. BTBT rate will increase when the biomolecule is immobilized.

5.2.4.2 Sensitivity Analysis of the TFET Device

The change in the drain current is due to the introduction of the biomolecule into the NC. Also, SIG-DP TEFT's current change rate is much more than that of the TFET structure. Noise immunity is increased due to the increase in the drain current, which is desired and promising. The SIG-DP TFET biosensor has the highest Sbio compared with the traditional TFET and DP TFET, which was required in this TEFT to be novel. Even though DP TFET has a high current while detecting the biomolecule, sensitivity is still a problem. The sensitivity will increase with the increase in the Ge composition. Before hybridization, the drain current will be high, leading to the decrease of S_{bio}.

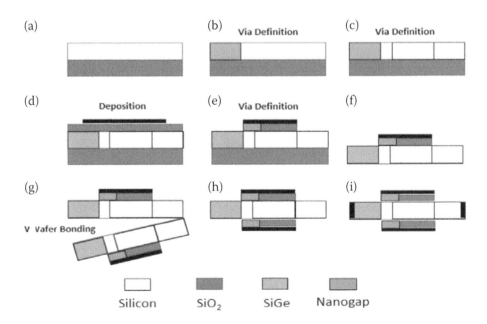

FIGURE 5.10 Fabrication of the TFET device.

5.3 CONCLUSION

Different biosensor devices have been investigated in this chapter after the careful comparison of these other TFET device structures and performance using simulation tools. Due to the band to band tunneling (BTBT) conduction mechanism, the TFET device overcomes the limitation and reduces the short channel effect. Hence, a TFET-based biosensor has developed as a suitable candidate for superior sensitivity and response time than a FET-based biosensor. In this chapter, the structure of the device and the parameters are investigated to design the TFET sensor devices. Hence, it will be more beneficial to create novel TFET devices for biosensor applications. The performance results show that the dielectric constant becomes constant, the thickness of the cavity grows smaller, the more positively charged, and the greater the sensitivity of the TFET device. Furthermore, simulation results show that the TFET structures can be applied for ultra-sensitive and low-consumption biosensor devices.

REFERENCES

[1] Barbaro M, Bonfglio A, Rafo L (2006). A charge-modulated FET for detection of biomolecular processes: conception, modeling, and simulation. *IEEE Transactions on Electron Devices*, 1, 158–166.

[2] Kannan N, Kumar MJ (2013). Dielectric-modulated impact-ionization MOS transistor as a label-free biosensor. *IEEE Electron Device Letters*, 34(12), 1575–1577.

[3] Jeon K (2010). Si tunnel transistors with a novel silicided source and 46mV/dec swing. In: Proceeding of the VLSI, 121–122.

[4] Sarkar D, Banerjee K (2012). Fundamental limitations of conventional-FET biosensors: quantum-mechanical-tunneling to the rescue. In: Proceedings of the IEEE device research conference, 83–84.

[5] Manoharan, AK, Chinnathambi, S, Jayavel, R, Hanagata, N (2017). Simplified detection of the hybridized DNA using a graphene field effect transistor. *Science and Technology of Advanced Materials*, 18, pp. 43–50.

[6] Sarkar D, Banerjee K (2012). Proposal for tunnel-field effect-transistors Ultra-sensitive and label-free biosensors. *Applied Physics Letters*, 100(14), 143108.

[7] Narang R, Reddy KS, Saxena M, Gupta RS, Gupta M (2012). A dielectric Modulated tunnel-FETbased biosensor for label-free detection: Analytical modeling study and sensitivity analysis. *IEEE Transactions on Electron Devices*, 59(10), 2809–2817.

[8] Kumar S, Singh Y, Singh B, Tiwari PK (2020). Simulation study of dielectricModulated dual channel trench gate TFET-based biosensor. *IEEE Sensors Journal*, J20(21), 12565–12573.

[9] Anam A, Anand S, Amin SI (2020). Design and performance analysis of Tunnelfield effect transistor with buried strained Si 1–x Ge x source Structure basedbiosensor for sensitivity enhancement. *IEEE Sensors Journal*, 20(22), 13178–13185.

[10] Chong C, Liu H, Wang S, Chen S (2021). Simulation and performance analysis of dielectric modulated dual source trench gate TFET biosensor. *Nanoscale Research Letters*, 16, 34.

[11] Wadhwa G, Raj B (2021). Surface potential modeling and simulation analysis of dopingless TFET biosensor. *Silicon*.

[12] Moon DI, Han JW, Meyyappan M (2016). Comparative study of feldefecttransistor based biosensors. *IEEE Transactions on Nanotechnology*, 15(6), 956–961.

[13] A dielectric-modulated field effect transistor for biosensing HYUNGSOON IM, XING-JIU HUANG, BONSANG GU AND YANG-KYU CHOI* Nano-Oriented Bio-Electronic Lab, School of Electrical Engineering and Computer Science, Korea Advanced Institute of Science and Technology, Daejeon 305-701, Republic of Korea.

[14] Stem E, Klemic JF, Routenberg DC, Wyiembak PN, Turner-Evazis DB, Hamilton AD, LaVan DA, Fahmy TM, Reed MA (2007) Label-free immune detection with CMOS-compatible semiconducting nano- wires. *Nature*, 445(7127), 519–522.

[15] Vaddi R, Agarwal RP, Dasguptn S (2011). Analytical modeling of subthrcshold current and subtlircshold swing of an underlap DGMOSF IT with tied—independent gate and symmetric—asymmetric options. *Microelectronics Journal*, 42(5), 798–807.

[16] Inn H, Huang X-J, Gu B, Choi Y-K (2007). A dielectric-modulated field-e8ect transistor for biosensing. *Nature Nanotechnology*, 2(7), 430–434.

[17] Xu P, Lou I-1, Zhang L, Yu Z, Lin X (2017). Compact model for double-gate tunnel FETs with gate-drain Underlap. *IEEE Transactions on Electron Devices,*, 64(12), 5242–5248.

[18] Gholizadeh M, Hosseini SE (2014). A 2-D analytical model for double gate tunnel FETS. *IEEE Transactions on Electron Devices*, 61(5).

[19] Ionescu AM and Riel H (2011). Tunnel field effect transistors as energy-efficient electronic switches. *Nature*, 479, 329–337.

[20] Boucart K and Ionescu AM (2007). Double-gate tunnel FET with high-κ gate dielectric. *IEEE Transactions on Electron Devices*, 54, 1725–1733.

[21] Kim CH, Jung C, Park HG and Choi YK (2008).Novel dielectric-modulated field effect transistor for label-free DNA detection. *BioChip Journal*, 2, 127–134.

[22] Tang X et al. (2009). Direct protein detection with a nano-interdigitated array gate MOSFET. *Biosensors and Bioelectronics*, 24, 3531–3537.

[23] Kim S, Baek D, Kim J-Y, Choi S-J, Seol M-L and Choi Y-K (2012). A transistor-based biosensor for the extraction of physical properties from biomolecules. *Applied Physics Letters*, 101, 073703.

[24] Shockley W and Read WT (1952). Statistics of the recombinations of holes and electrons. *Physical Review*, 87, 835–842.

[25] Kannan N and Kumar MJ (2013). Dielectric-modulated impact-ionization MOS transistor as a label-free biosensor. *IEEE Electron Device Letters*, 34, 1575–1577.

[26] Parihar MS and Kranti A (2015). Enhanced sensitivity of double gate junctionless transistor architecture for biosensing applications. *Nanotechnology*, 26, 145201.

[27] Im H, Huang X-J, Gu B and Choi Y-K (2007). A dielectric-modulated field effect transistor for biosensing. *Nature Nanotechnology*, 2, 430–434.

[28] Narang R, Saxena M, and Gupta M (2015). "Comparative analysis of dielectric-modulated FET and TFET-based biosensor," *IEEE Transactions on Nanotechnology*, 14(3), 427–435.

[29] Anvarifard MK, Ramezani Z, and Amiri IS (2019). "Proposal of an Embedded Nanogap Biosensor by a Graphene Nanoribbon Field-Effect Transistor for Biological Samples Detection," *Physica status solidi (a)*, 217(2), 1900879.

[30] Narang R, Reddy KS, Saxena M, Gupta R, and Gupta M (2012). "A dielectric-modulated tunnel-FET-based biosensor for label-free detection: Analytical modeling study and sensitivity analysis," *IEEE Transactions on electron devices*, 59(10), 2809–2817.

[31] Chivers CE, Koner AL, Lowe ED, and Howarth M (2011). "How the biotin–streptavidin interaction was made even stronger: investigation via crystallography and a chimaeric tetramer," *Biochemical Journal*, 435(1), 55–63.

[32] SILVACO ATLAS User's Manual: 2-D Device Simulator, Santa Clara, CA, USA, 2012 https://silvaco.com/tcad/victory-device-3d/
[33] Singh S, Raj B (2020). "Analytical Modeling and Simulation analysis of T-shaped III-V heterojunction Vertical T-FET", *Superlattices and Microstructures*, Elsevier, 147, 106717, Nov.
[34] Chawla T, Khosla M, Raj B (2020). "Optimization of Double-gate Dual material GeOI-Vertical TFET for VLSI Circuit Design, *IEEE VLSI Circuits and Systems Letter*, 6(2), 13–25, Aug.
[35] Kaur M, Gupta N, Kumar S, Raj B, Singh AK (2020). "RF Performance Analysis of Intercalated Graphene Nanoribbon Based Global Level Interconnects" *Journal of Computational Electronics, Springer*, 19, 1002–1013, June.
[36] Wadhwa G, Raj B (2020). "An Analytical Modeling of Charge Plasma based tunnel field effect transistor with Impacts of Gate underlap Region" *Superlattices and Microstructures*, Elsevier, 142, 106512, June.
[37] Singh S, Raj B (2020). "Modeling and Simulation analysis of SiGe hetrojunction Double GateVertical t-shaped Tunnel FET", *Superlattices and Microstructures*, Elsevier, 142, 106496, June.
[38] Singh S, Raj B (2020). "A 2-D Analytical Surface Potential and Drain current Modeling of Double Gate Vertical t-shaped Tunnel FET", *Journal of Computational Electronics, Springer*, 19, 1154–1163, Apl.
[39] Singh S, Bala S, Raj B, Raj Br (2020). "Improved Sensitivity of Dielectric Modulated Junctionless Transistor for Nanoscale Biosensor Design", *Sensor Letter, ASP*, 18, 328–333, Apl.
[40] Kumar V, Kumar S and Raj B (2020). "Design and Performance Analysis of ASIC for IoT Applications" *Sensor Letter, ASP*, 18, 31–38, Jan.
[41] Wadhwa G, Raj B (2019). "Design and Performance Analysis of Junctionless TFET Biosensor for high sensitivity" *IEEE Nanotechnology*, 18, 567–574.
[42] Wadhera T, Kakkar D, Wadhwa G, Raj B (2019). "Recent Advances and Progress in Development of the Field Effect Transistor Biosensor: A Review" *Journal of Electronic Materials*, Springer, 48(12), 7635–7646, December.
[43] Singh S, Raj B (2019). "Design and analysis of hetrojunction Vertical T-shaped tunnel field effect transistor", *Journal of Electronics Material, Springer*, 48(10), 6253–6260, October.
[44] Goyal C, Ubhi JS and Raj B (2019). "A Low Leakage CNTFET based Inexact Full Adder for Low Power Image Processing Applications", *International Journal of Circuit Theory and Applications*, Wiley, 47(9), 1446–1458, September.
[45] Sharma, SK, Raj, B, Khosla, M (2019). "Enhanced Photosensivity of Highly Spectrum Selective Cylindrical Gate In1-xGaxAs Nanowire MOSFET Photodetector", *Modern Physics Letter-B*, 33(12), 1950144.
[46] Singh J, Raj B (2019). "Design and Investigation of 7T2M NVSARM with Enhanced Stability and Temperature Impact on Store/Restore Energy", *IEEE Transactions on Very Large Scale Integration Systems*, 27(6), 1322–1328, June.
[47] Bhardwaj AK, Gupta S, Raj B, Singh A (2019). "Impact of Double Gate Geometry on the Performance of Carbon Nanotube Field Effect Transistor Structures for Low Power Digital Design", *Computational and Theoretical Nanoscience*, ASP, 16, 1813–1820.
[48] Goyal C, SUbhi J and Raj B (2019). Low Leakage Zero Ground Noise Nanoscale Full Adder using Source Biasing Technique, *"Journal of Nanoelectronics and Optoelectronics"*, American Scientific Publishers, 14, 360–370, March.
[49] Singh A, Khosla M, Raj B (2019). "Design and Analysis of Dynamically Configurable Electrostatic Doped Carbon Nanotube Tunnel FET", *Microelectronics Journal*, Elsevier, 85, 17–24, March.

[50] C Goyal, JS Ubhi and B Raj (2018). A reliable leakage reduction technique for approximate full adder with reduced ground bounce noise, *Journal of Mathematical Problems in Engineering, Hindawi*, 2018, Article ID 3501041, 16 pages, 15 Oct.

[51] Wadhwa G, Raj B (2018). "Label Free Detection of Biomolecules using Charge-Plasma-Based Gate Underlap Dielectric Modulated Junctionless TFET" *Journal of Electronic Materials (JEMS), Springer*, 47(8), 4683–4693, August.

[52] Wadhwa G, Raj B (2018). "Parametric Variation Analysis of Charge-Plasma-based Dielectric Modulated JLTFET for Biosensor Application" *IEEE Sensor Journal*, 18(15), AUGUST 1.

[54] Singh G, Sarin RK and Raj B (2018). "Fault-Tolerant Design and Analysis of Quantum-Dot Cellular Automata Based Circuits", *IEEE/IET Circuits, Devices & Systems*, 12, 638–664.

[53] Yadav D, Chouhan SS, Vishvakarma SK and Raj B (2018). "Application Specific Microcontroller Design for IoT based WSN", *Sensor Letter, ASP*, 16, 374–385, May.

[55] Singh J, Raj B (2018). "Modeling of Mean Barrier Height Levying Various Image Forces of Metal Insulator Metal Structure to Enhance the Performance of Conductive Filament Based Memristor Model", *IEEE Nanotechnology*, 17(2), 268–267, March (SCI).

[56] Jain A, Sharma S, Raj B (2018). "Analysis of Triple Metal Surrounding Gate (TM-SG) III-V Nanowire MOSFET for Photosensing Application", *Opto-electronics Journal, Elsevier*, 26(2), 141–148, May.

[57] Jain N, Raj B (2018). "Parasitic Capacitance and Resistance Model Development and Optimization of Raised Source/Drain SOI FinFET Structure for Analog Circuit Applications, *Journal of Nanoelectronics and Optoelectronins, ASP, USA*, 13, 531–539, Apl.

[58] Mani, N, Moh, M, & Moh, TS (2021). Defending deep learning models against adversarial attacks. *International Journal of Software Science and Computational Intelligence (IJSSCI)*, 13(1), 72–89.

[59] Sarivougioukas, J, & Vagelatos, A (2020). Modeling deep learning neural networks with denotational mathematics in UbiHealth environment. *International Journal of Software Science and Computational Intelligence (IJSSCI)*, 12(3), 14–27.

[60] Tewari, A, et al. (2020). Secure Timestamp-Based Mutual Authentication Protocol for IoT Devices Using RFID Tags. *International Journal on Semantic Web and Information Systems (IJSWIS)*, 16(3), 20–34.

[61] Dwivedi, RK, Kumar, R, & Buyya, R (2021). Secure healthcare monitoring sensor cloud with attribute-based elliptical curve cryptography. *International Journal of Cloud Applications and Computing (IJCAC)*, 11(3), 1–18.

[62] Mamta, Gupta, BB, Li, KC, Leung, VC, Psannis, KE, & Yamaguchi, S (2021). Blockchain-assisted secure fine-grained searchable encryption for a cloud-based healthcare cyber-physical system. *IEEE/CAA Journal of Automatica Sinica*, 8(12), 1877–1890.

[63] Sarrab, M, & Alshohoumi, F (2021). Assisted-fog-based framework for iot-based healthcare data preservation. *International Journal of Cloud Applications and Computing (IJCAC)*, 11(2), 1–16.

Security-Based Genetic Algorithms for Health Care

6

Neeraj Kumar Rathore
Associate Professor, Department of Computer Science, Indira Gandhi National Tribal University (A Central University), Amarkantak, M.P., India

Contents

6.1	Introduction	124
6.2	The Need of Security-Based Genetic Algorithms in Health Care	125
6.3	Basic Terminologies Related to Security-Based GA	125
	6.3.1 Chromosomes	126
	6.3.2 Populations	126
	6.3.3 Genes	126
	6.3.4 Alleles	127
	6.3.5 Genotypes and Phenotypes	127
6.4	General Genetic Algorithm for Health Care	127
6.5	Operators in GA for Health Care	129
	6.5.1 Encoding	129
	6.5.2 Selection	130
	6.5.3 Crossover	133
	6.5.4 Mutation	137
6.6	Stopping Condition for Genetic Algorithm for Health Care the Various Stopping Condition	138

DOI: 10.1201/9781003189633-6

Conclusion	143
Author's details	144
References	144

6.1 INTRODUCTION

Based on evolutionary ideas of natural selection and genetics, a catechistic searching technique has evolved, which is called the genetic algorithm. Here, heuristic means a technique designed to solve the problem quickly i.e. guarantee to give the best solution but the optimal solution may or may not be obtained. Security-based GA is focused on security and optimization, which means the process of making something better. In biology, we studied genes, chromosomes. Two parents create new offspring and pass their best to its next generation. The new generation then tries to create the best solution. Here, genes travel from one generation to the next generation. The security of GA is based on this idea, which is used in AI and ML where we have a lot of solutions and how to find optimal solutions among them.

In general, terms, if we have N number of solutions for resolving a specific problem that all the solutions cannot yield optimize solution of that problem but security-based GA assist in finding and optimizing the solution for all kinds of problems including security, whether it is constrained or unconstrained one. It emerges over time so that a reform solution can be achieved and searching can be made refined.

Charles Darwin's theory:

- Within any population, there exist natural variations. Some individuals exhibit more favorable variations than others.
- The struggle for survival eliminates the unfit individuals. The fit individuals processing favorable variation survive and reproduce. This is referred to as "natural selection" (or survival of the fittest).
- The individuals having favorable variations pass on these variations to their progeny (offspring) from generation to generation.

A security-based genetic algorithm has the potential to bear a "good enough" solution "fast enough." Evolution is known to be a successful robust method for adaptation with biological healthcare systems. Motivation to adopt security-based GA comes from the following given points:

- Solving difficult problems: Security-based GA demonstrate as a powerful tool to produce valuable near-optimal solutions in less period.
- Breakdown of gradient-based methods: In most of the real-world situations, where a very complex problem undergoes an implicit tendency of getting stuck at the local optima.

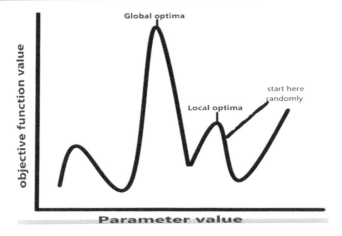

FIGURE 6.1 Position global optima and local optima.

- Obtaining a good solution fast: Few laborious problems such as Travelling Salesman Problem (TSP) possess practical applications such as pathfinding and VLSI design [1–15] (Figure 6.1).

6.2 THE NEED OF SECURITY-BASED GENETIC ALGORITHMS IN HEALTH CARE

- They are more robust than conventional algorithms.
- If the inputs are modified to some extent or within the presence of noise, security-based GA does not break easily, unlike in older AI systems.
- Security-based GA may exhibit compelling benefits over further usual optimization approach (e.g. linear programming, DFS, BFS, heuristic, etc.) when searching in a large state-space, multimodal state-space, or n-dimensional surface.

6.3 BASIC TERMINOLOGIES RELATED TO SECURITY-BASED GA

In a biological sense, the smallest unit of life is a human/animal cell. The cell nucleus can be demonstrated as the center of the cell. The information related to genetics stored in the cell nucleus. In the nucleus, there is a nuclear envelope; within this, there is a nucleolus and set of chromosomes. Chromosomes get a hold onto all the genetic data. Every chromosome is built up of DNA (deoxyribonucleic acid). Gene can be stated as the division of several parts of chromosomes. Alleles can is demonstrated as the plausibility of gene aggregation for one property and distinct alleles constitute a gene. For instance, consider a gene for eye color and the various prospects of alleles are brown, black, green, and blue. A gene pool can be stated as the set of all the probable alleles incorporate in a population. The gene pool can regulate all the probable divergence for upcoming generations.

A genotype can be stated as the complete aggregation of genes for a selected individual. The substantial aspect of decoding a genotype can be signified as a phenotype. The reproduction recombines genotype and the choice is always done on phenotype [10–25].

6.3.1 Chromosomes

To figure out a specific problem, in general, if we have N number of solutions. So, one such solution to a given problem is a chromosome (Figure 6.2).

FIGURE 6.2 Representation of chromosome.   1010101110101101

6.3.2 Populations

A population is a collection of individuals (chromosomes). That is a subgroup of total solutions to the given problem. There are two main aspects of the population used in security-based GA in health care:

a. Initial population generation: In utmost cases, it is selected randomly. Typically, for provoking the initial population, a sort of catechistic can be used.
b. Population size: It is relying upon the complexity of the problem. The random initialization of the population can be typically employed for population size. Every bit is initialized to either 0 or 1 in binary encoded chromosomes.

6.3.3 Genes

The sequence of genes constitutes a chromosome. We can also say that one element position of the chromosome (Figure 6.3).

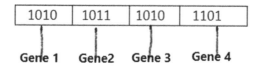

FIGURE 6.3 Representation of gene.

6.3.4 Alleles

For a particular chromosome, a gene takes a value that is called an allele (Figure 6.4).

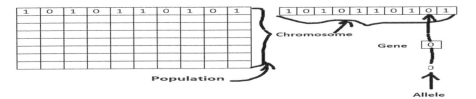

FIGURE 6.4 Representation of allele, gene, chromosome, and population concept.

6.3.5 Genotypes and Phenotypes

A phenotype can be defined as the population in the actual real-world solution space i.e. row solution without proper representation in writing.

But when we give input in the algorithm, proper representation is required, and for that we use a genotype. The population in the computational space is termed a genotype.

Decoding can be stated as the method that involves remodeling a solution from the genotype to the phenotype space. Encoding is the method that involves remodeling a solution from the phenotype to genotype space (Figure 6.5).

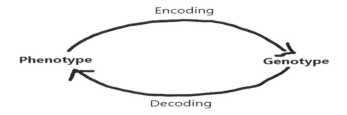

FIGURE 6.5 Representation of encoding and decoding.

6.4 GENERAL GENETIC ALGORITHM FOR HEALTH CARE

The general procedure of the security-based genetic algorithm [25–34] is shown in Figure 6.6:

Step 1: Initial population
Initial population means randomly selecting solutions among all the possible solutions. The important thing is diversity, which means selected solutions do not focus on particular points; they should focus on

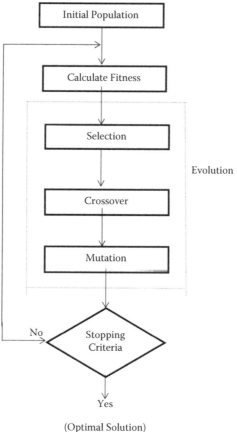

FIGURE 6.6 Flowchart for GA.

diverse points. As much as our input is diverse, the algorithm will properly work.

Step 2: Calculate fitness function

Fitness function is attached to each individual in the initial population. It is a function you want to optimize or we can say that function that takes solution as input and produce more relevant (stable) output.

Step 3: Selection

Selection is directly proportional to the fitness function, which means selecting that promising solution whose fitness function value is more.

Step 4: Crossover

In biological terms, combining two parents to generate children for the upcoming generations is a chromosome. In other term, we can say that in the crossover, two-parent solutions are taken into account and produce a new population so that its fitness value is more. Crossover can be performed by various techniques like the one-point crossover, multipoint crossover, etc.

Step 5: Mutation

The new population (strings) generated by the crossover process is subjected to mutation. It is a small random tweak in the chromosome to get a new solution. A mutation can be of different forms like flipping, interchanging, reversing, etc. A mutation maintains the diversity of the population and prevents the algorithm to be trapped in a local minimum. It incorporates riffling a bit from 1 to 0 and vice versa.

Step 6: Stopping criteria

Stopping criteria means the convergent point where the fitness value is highest. It means all the newly generated populations give optimal solutions. If there is a condition where optimal solutions do not produce, then the whole process is repeated until we reach the convergent state or endpoint.

6.5 OPERATORS IN GA FOR HEALTH CARE

The fundamental operators used in the security-based genetic algorithm are encoding, selection, crossover, and mutation. These operators and their types are discussed below.

6.5.1 Encoding

The process that represents specific genes is called encoding. Encoding can be carried out by employing bits, numbers, trees, arrays, lists, or any other objects.

- Binary Encoding: It is the most common way of encoding. Each chromosome encodes in a binary (bit 0s and 1s) string. Every bit in a string represents few characteristics of the solution. Here, string length depends on accuracy (Figure 6.7).

Chromosome 1	1 0 1 0 0 0 0 1 0 0 0 0
Chromosome 2	0 1 1 0 1 1 0 1 1 0 1 0

FIGURE 6.7 Binary encoding.

- Octal Encoding: In this type of encoding scheme, octal numbers (0–7) constitute a string (Figure 6.8).

Chromosome 1	02466237
Chromosome 2	16315661

FIGURE 6.8 Octal encoding.

- Hexadecimal Encoding: Hexadecimal numbers (0–9, A–F) are employed for encoding strings in this scheme (Figure 6.9).

Chromosome 1	8BE9
Chromosome 2	4FAC

FIGURE 6.9 Hexadecimal encoding.

- Permutation Encoding: It is also known as real number encoding. In this encoding, every chromosome is a string of integer/real values, which represent a number in a sequence. This type of encoding is only useful for ordering problems (Figure 6.10).

Chromosome 1	4 5 3 2 6 1 7 9 8
Chromosome 2	8 9 7 1 6 2 3 4 5

FIGURE 6.10 Permutation encoding scheme.

- Value Encoding: In this type of encoding, each chromosome consists of a string of values. Here, values that are associated with the problem may be in the mode of real numbers, numbers, or characters for few sophisticated objects. This type of scheme serves as a decent choice for specific issues wherever a different variety of encoding would be troublesome (Figure 6.11).

Chromosome A	4.3234 0.1234 5.5446 1.9293 6.5454
Chromosome B	ABDJEIFJDHDIERTUGNMOPLDGT
Chromosome C	(right), (back), (forward), (left), (right)

FIGURE 6.11 Value encoding scheme.

- Tree Encoding: Genetic programming employed this type of encoding scheme for emerging program expression. Here, each chromosome is represented in the form of a tree of a few objects and these objects can be functions and commands used in programming language [21–32].

6.5.2 Selection

In this technique, two parents are selected from the population for crossing. The consequent step once agreeing on encoding is to find out a way to select individuals within a population that may turn out descendent for the consequent generation and the way through which several descendants will each individual generate. The choice aims to intensify in hops competent individuals within the population that their descendants are more fit. Chromosomes are chosen to be reproductive parents from the initial population. The main focus is on the way to select chromosomes. The strongest ones live for producing new descendents, as per Darwin's theory of evolution.

According to the function of evaluation, the selection process randomly selects chromosomes within the population. The higher the fitness level, the greater the chance of being chosen as an individual. The extent to which the most appropriate individuals are preferred is termed *selection pressure*. As the selection pressure is high, it means more preference is granted to the best individuals.

In health care, the security-based genetic algorithm convergence rate is primarily dictated by the selection pressure level, with higher selection pressures resulting in higher convergence rates. Genetic algorithms are capable of classifying optimal or almost optimal solutions under a wide variety of selection scheme pressure. A security-based genetic algorithms can take an unnecessarily long time to search out optimization solutions if the convergence rate is slow because of too low of a selection pressure.

Prematurely converging to an incorrect solution can result in an increased security-based genetic algorithm shift is happening because of too high a selection pressure. From preventing premature convergence, in conjunction with the selection constraints, the distinction of populations should also retain by the selection scheme.

Usually, there are two categories of selection: system-proportionate selection based and ordinal selection based. In the first case, individuals can be eliminated based on their fitness value as compared to the fitness of some other individuals within the population. In ordinal selection schemes, individuals are chosen not based on their raw fitness but in keeping with their ranking in the population. This shows the independence of selection pressure on population fitness distribution, and this focused exclusively on population ranking.

A scaling function may also be used to reconstruct the population's fitness range, to regulate the selection pressure. As an illustration, if the entire solution has its fitness within the range [888,999], the chance of choosing a prominent individual as compared to that employing a proportionate-based approach would not be significant. Once every individual's fitness range equal to that of [0,1], the chance of choosing a better individual rather than a poor one would be essential.

Selection needs to be stabilized with mutation variability and crossover variability. Sub-optimal and extremely fit individuals can seize the population if the too strong selection is carried out, which decreases the heterogeneity required for amendment and growth; too slow progression results if the too slow selection is carried out. There are numerous strategies of selection that are given below:

   a. Roulette Wheel Selection: Roulette wheel selection is one of the popular methods used for the selection method of security-based genetic algorithms. Here, the proportionate selection operator is used. This method is often a comparatively robust selection strategy, as fit individuals don't seem to be expected to be picked for; however, are far more possible to be. In roulette selection theory, the roulette wheel is used by which a linear search is performed with the slots weighted in proportionate to the individual's fitness value by setting a target value that is given by the random proportion of the population's number of finesses. Until the population reaches the target value, it is moved through.

   This methodology can be explained below in a stepwise fashion:

i. Let T indicate summation of the entire value of population within the population.
ii. The following process is repeated N numbers of time:
   - Select any random integer number "r" that comes within the range 0 and T.
   - Until this sum is greater than or equal to "r," loop through the individuals in the population and summing the expected value continues. That individual is selected whose sum of expected value is greater than this limit.

   The main advantage of this method is that it is easy to implement and the drawback is that it is noisy.

b. Random selection: As the name indicates, the random selection of a parent from the population is involved. On average, random selection is a little more tedious compared to the roulette wheel selection method, in the sense of interruption of genetic codes.

c. Rank selection: The roulette wheel would have an issue when the fitness value is very different, as in this case, the chromosomes whose fitness value is very low, fewer chances of selection. The rank selection method involved ranks the population and each chromosome from the ranking perceive fitness. Fitness 1 indicates the worst and fitness N means the best chromosome. This leads to gradual convergence but avoids convergence too quickly. It retains diversity and, thus, results in a favorable search. Capable parents are chosen in turn.

d. Tournament selection: In this selection strategy, based on fitness value, a tournament is carried out between N individuals in an N-way tournament selection. So, in this way, only the fittest individual is selected and passed to the successor generation. We get our final selection of individuals by performing many such tournaments and these passes to the next generation. One who has the highest fitness value is selected as a winner of the tournament. The selection pressure is given by the gap in fitness value, whose advantage is that it pushes the security-based genetic algorithm for generating the fittest successor gene. This technique is known to be more powerful and able to identify an optimal solution.

e. Boltzmann selection: Here, the selection rate is monitored by a frequently fluctuating temperature. Here, selection pressure is negatively correlated to the temperature. Initially, the temperature is kept high, so the selection pressure is small as per their relation with temperature. Gradually, as the temperature is dropped, the selection pressure increases. As a result, search space becomes contracting in conjunction with the preservation of population divergence. The Boltzmann probability for individual selection is determined as:

$$P = \exp -\frac{[f_{max} - f(X_i)]}{T} \tag{6.1}$$

Here, $T = T_o(1 - \alpha)^k$ and $k = (1 + 100 * g/G)$; g = current generation number, G = maximum value of g.

In this selection strategy, the probability is very high of choosing the best string formatting. Also, this method incorporates a very small execution time. One of the drawbacks of this method is that during the mutation stage some information may have vanished. But this can be avoided by elitism.

Elitism is a method that consists of the inclusion of the first prime chromosome or some prime chromosomes are transcribed to the new population.

f. Stochastic universal sampling: This type of sampling method produces zero bias and minimal distribution. Here, the individual is mapped to coterminous fragments of the line, and like roulette-wheel selection, every fragment of its individual is equal in size to its fitness. As much as individuals are to be selected, in this method evenly spaced pointers are put along a line. Let the number of individuals to be selected is given by N pointer, then the distance between the pointers is given by 1/N pointer and the range [0, 1/NPointer] indicates the location of the first pointer, which can be a randomly generated number.

The stochastic universal sampling concept is easily understood by Figure 6.12. The distance between the pointers is 1/6 = 0.167 for the selection of six individuals. As a sample, one of the random numbers to be chosen within the range [0, 0.125] is 0.1.

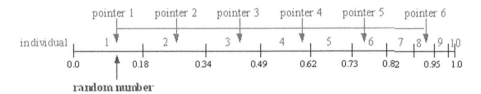

FIGURE 6.12 Stochastic universal sampling.

The mating population consists of the individuals after the selection process is given by 1, 2, 3, 4, 6, 8.

As compared to roulette wheel selection, stochastic universal sampling ensures a selection of offspring which is closer to what is deserved.

6.5.3 Crossover

Security-based GA's basic technique for producing the best solutions/offspring is to cross over the parent genes. Optimal solution in minimal different crossover strategies are designed to get the generations as early as possible. Crossover operator selection has more impact on security-based GA's efficiency. A selection of suitable breeding operators may prevent premature convergence in GA.

- Single-Point Crossover
  For the parent organism, a crossover point is chosen on the string. In the string of species, all the data that are beyond that stage are exchanged between the two parent string species and are labeled with positional bias.

Chromosome1	11011|00100110110
Chromosome2	11011|11000011110
Offspring1	11011|11000011110
Offspring2	11011|00100110110

Single Point Crossover

- Two-Point Crossover
  This type of crossover technique is a special case by N-point. On the individual chromosomes (strings), and the genetic material is exchanged at the points where random points are chosen

Chromosome1	11011|00100|110110
Chromosome2	10101|11000|011110
Offspring1	11011|11000|110110
Offspring2	10101|00100|011110

Two Point Crossover

- Multi-Point Crossover
  This crossover has two ways to go: even number of cross-sites and odd number of cross-sites. Even the number of cross-sites involves randomly chosen cross-sites surrounding a circle and information is swapped along those cross-sites. The odd number of cross-sites involves at the beginning of the series a separate cross point is often assumed.
- Uniform Crossover
  The chromosomes are not divided into fragments in a uniform fusion; rather, they handle every gene on an individual basis. To check for every chromosome enclosed within the offspring or not, we tend to primarily flip a coin for every chromosome. We tend to skew the coin to one parent, too, to possess a lot of genetic material from that parent within the infant (Figure 6.13).

Parent 1	1 0 1 0 1 0 10
Parent 2	0 1 0 1 0 1 01

Child 1	1 1 0 0 1 0 01
Child 2	0 0 1 1 0 1 10

FIGURE 6.13 Uniform crossover.

- Three-Parent Crossover
  Three parents are selected at random in this crossover technique. The first parent's every bit is contrasted with the second parent's bit. If all are the same, the offspring takes the bit; otherwise, the offspring takes the third parent's part (Figure 6.14).

Parent 1	0 1 0 1 0 0 0 1
Parent 2	1 0 0 0 1 0 01
Parent 3	0 1 0 1 0 1 0 1
Child	0 1 0 1 0 0 0 1

FIGURE 6.14 Three-parent crossover.

- Crossover with Reduced Surrogate
  The reduced surrogate operator constrains crossover to create new individuals whenever possible. This is achieved by limiting the position of the crossover point, so that crossover points occur only where the gene value varies.
- Shuffle Crossover
  Shuffle crossover has to do with standard crossover. A single location is picked on the crossover. But in both parents, they're randomly shuffled until the variables are exchanged. The variables within the offspring are unshuffled after recombination. This eliminates positional bias as the variables are reassigned at random every time the crossover is performed.
- Precedence Preservative Crossover
  PPX was developed autonomously for vehicle routing problems by Blanton and Wainwright (1993), and problems with scheduling by Bierwirth et al. (2008). The operator passes priority operations relationships given in two parental permutations for one offspring at the same rate, and no new precedent relationships are set in motion. PPX is shown below for an issue comprised of six operations A–F. The operator functions as follows:
  ○ The length vector Sigma, sub i = 1 to mi, which represents the number is randomly filled with operations involved in the problem place Elements {1, 2}.
  ○ This vector sets the order in which the operations are carried outdrawn successively from parent 1 and parent 2.
  ○ The operations "append" and "delete" can also be considering permutations for parents and offspring as lists.
  ○ It can be initiate by initializing an empty offspring first.
  ○ One among two parents has the leftmost operation selected according to the parent order given in the vector.
  ○ Once an operation has been identified, all parents uninstall it.
  ○ The preferred operation is eventually appended to the offspring.
  ○ Until both parents are zero and the child incorporates all relevant operations, repeat the preceding steps.
  ○ It should be noted that due to the "delete append" scheme, the PPX scheme is not working in a standardized crossover manner (Figure 6.15).

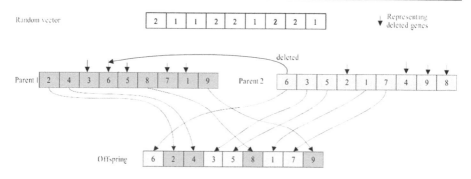

FIGURE 6.15 Precedence preservative crossover.

- Ordered Crossover
  Whenever the problem is order-based mostly, an ordered two-point crossover is employed. Based on two-parent chromosomes, two random crossover points are chosen and partitioning into three parts 1.e. left part, middle part, and right part. This ordered two-point crossover acts as follows: child 1 inherits its left and right part from the left and right part of its parent and the child middle part is decided by the genes in parent 1's middle part within the order during which the values seem in parent 2. In determining child 2, an analogous process is applied (Figure 6.16).

    Parent 1: 4 3 |1 2| 5 6   Child 1: 4|3 2|15 6

    Parent 2: 3 2| 1 7| 6 5   Child 2: 3| 2 7|1 65

FIGURE 6.16 Ordered crossover.

- Partially Matched Crossover
  Travelling Salesman Problem (TSP) is one of the applications in which PMX can be applied. TSP chromosomes are represented by the integer's sequences, where each integer serves as a different city and the order serves as the time a city is visited at. Here, the only point of concern is permutation encoding, which is the labels under this representation. It can be interpreted as a crossover of perm tations that ensures that all roles are located exactly once in each offspring, i.e. all offspring obtain a full complement of genes, followed by their parents' reciprocal filling of alleles. PMX operates as follows:
  i. The two selected chromosomes align themselves.
  ii. As per the section specified to suit, randomly pick two locations along a string.
  iii. A cross-position exchange operation is performed through position-by-position, where the matching section is employed.
  iv. The movement of alleles into their new position is involved in their child.

The working steps of PMX are depicted as follows [31-24]:

- As an example, two strings are taken into account in Fig. 3.14
- Wherever the selected cross points are represented by dots.
- The position-wise exchange that has to occur in each parent to produce offspring described in a subsequent step.
- Scan the swapping from the identical section of a chromosome on any chromosome.
- In the example, the place-exchange numbers are 5 and 3, 6 and 10, and 7 and 2. The resulting offspring are shown below (Figure 6.17).

Name 9 8 4 . 5 6 7 . 1 3 2 10   Allele1 0 1 . 0 0 1 . 1 1 0 0

Name 8 7 1 . 2 3 10 . 9 5 4 6   Allele 1 1 1 . 0 1 1 . 1 1 0 1

Given Strings

Name 9 8 4 . 3 10 2 . 1 5 7 6 Allele 1 0 1 . 1 0 0 . 1 0 10

Name 8 2 1 . 7 5 6 . 9 3 4 10 Allele1 0 1 . 1 1 1 . 1 1 0 1

FIGURE 6.17 PMX crossover.

6.5.4 Mutation

The strings endure mutations by the following crossover. A mutation prohibits the algorithm from being captured in a minimum of the locality. The main role of mutation is to retrieve the vanished genetic materials in conjunction with the randomly disturb genetic information. Here, mutation of a bit means flipping a bit, dynamical from 0 to 1 and vice versa.

- Flipping
  Based on a generated mutation chromosome, flipping of a bit involves changing 0 to 1 and 1 to 0. A parent is considered and a mutation chromosome is randomly generated. The child chromosome is produced by flipping a corresponding parent chromosome for a 1 in mutation chromosome. Binary encoding generally used the concept of flipping (Figure 6.18).

Parent	0 1 0 10 1 10
Mutation Chromosome	0 0 1 0 1 0 1 1
Child	0 1 1 1 1 1 0 1

FIGURE 6.18 Mutation flipping.

- Interchanging
  The bits analogous in those positions are interchanged, where the randomly chosen position of a string is involved (Figure 6.19).

FIGURE 6.19 Interchanging.

Parent	1 1 1 0 0 0 0 1
Child	1 1 0 0 0 1 0 1

- Reversing

  In case a reversing mutation is applied for a binary-encoded chromosome, the random position is chosen and the bits next to that position are reversed and the child chromosome is produced (Figure 6.20).

FIGURE 6.20 Reversing.

Parent	1 0 0 0 1 0 1 0
Child	1 0 0 0 1 0 0 1

- Mutation Probability

  How many portions of the chromosome are mutated is determined by the probability of mutation. The offspring is produced without any modification immediately after the crossover if there is no mutation process involved. One or more parts of the chromosome will be affected or modified if the mutation process happens. If the chance of mutation to be happening is 100 percent, the entire chromosome will change; if it is 0 percent, no change will occur. The major advantage of a mutation is that it prevents a security-based GA from the extremes of the locality.

6.6 STOPPING CONDITION FOR GENETIC ALGORITHM FOR HEALTH CARE THE VARIOUS STOPPING CONDITION

1. Maximum generations:

   When the specified number of generations has evolved, security-based GA stops at that point.
2. Elapsed time:

   The genetic process will end at the point where the maximum number of generations has been reached before the specified time has evolved.
3. No change in fitness:

   The genetic process will end at the point where the maximum number of generations has been reached before the specified number of generations with no change in the population's best fitness.
4. Stall generation:

   Whenever the objective function does not improve for a sequence of successive generations of length "stall generation," the security-based GA ends at this point.
5. Stall time limit:

   If the situation occurs in which objective function does not improve during the period (in seconds) which is equal to "stall time limit," the security-based GA stops at this point.

The culmination of the convergence criterion ultimately puts a halt to the search. Some methods of culmination techniques are given below:
a. Best Individual:
   If the minimum fitness in the population falls down the convergence value, this criterion stops the search. This will leads the search to a better and rapid conclusion, ensuring a minimum favorable solution.
b. Worst Individual:
   It finishes the search, whenever the least fit individuals in the population have fitness lower than the convergence criterion. It ensures a minimum norm for the whole population. In any situation, a strict criterion will never be reached, in which situation the search will stop after the limit has been surpassed.
c. Sum of Fitness:
   This scheme is taken into account whenever the sum of fitness in the whole population not exceeding to the convergence value in the population record. This parameter ensures that all the individuals in the population are within a given fitness range. On setting the convergence value, consideration must be given to population size.
d. Median Fitness:
   If at least one-half of the individuals should be better than or equal to the convergence value, there is a great extent of choices are available in this criteria.
6. Security-based GA application for health care

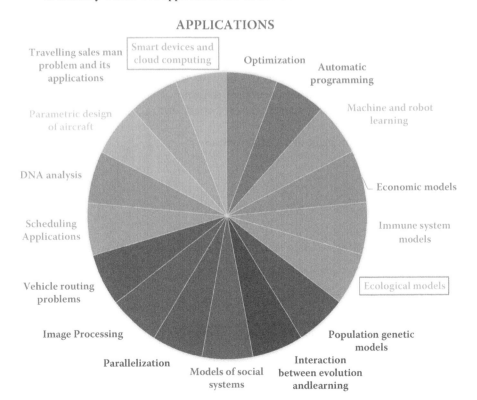

7. Constraints in genetic algorithm for health care

In unconstrained optimization problems, security-based GA on security is considered to comprise solely an objective function and no knowledge regarding the variable specification. An example of an unconstrained optimization problem can be given as

$$\text{Minimize } f(x) = x^2 \tag{6.2}$$

Here, no information is available regarding the "x" range. By using the random specification of its operators, security-based GA minimizes this function.

In constrained optimization problems, the information regarding variable specification will be given. In these problems, constraints are listed below:

a. Equality relations
b. Inequality relations

The series of parameters to be evaluated by security-based GA using the system being considered, the objective function (to be minimized or maximized), and constraints. However, the goal function is evaluated on running the program and the constraints are tested to see if there are any breaches. When no violations occur, the fitness value analogous to the objective function measurement is assigned to the specified parameter. The remedy is unreliable and results in no fitness in the case of constraint violation. As a consequence, some knowledge should be derived from the infeasible solution, regardless of their fitness rating in correlation to the extent of the breach of the restriction. With the help of the penalty method, we can achieve this.

By using a penalty approach that involves the transformation of a problem of constraint optimization into an unconstrained problem of optimization by allying a penalty or expense with entire combinations of constraints. Assessment of objective function embedded such type of penalty.

Let us examine a problem under constraint as

$$\text{Maximize } f(x) \tag{6.3}$$

Subject to $g_i(x) \geq 0$, $i = 1, 2, 3, \ldots, n$

where, x = k-vector. Transforming this to an unconstrained form is

$$\text{Maximize } f(x) + P \sum_{i=0}^{n} \emptyset \, [gi(x)] \tag{6.4}$$

where, $\emptyset$ = penalty function and P = penalty coefficient.

When the penalty coefficient reached infinity, the unconstrained solution transformed to the constrained solution.

8. Problem-solving using GA in health care

Problem statement: Maximize the function $f(x) = x^2$ with x lying in interval [0, 31] i.e. x = 0, 1, 2, ..., 31.

1. Randomly generate initial population. They are called chromosomes and genotypes.
   Here, for example, four strings of 5-bit length taken as chromosomes are: 01101 (13), 11000 (24), 01000 (8), 10011 (19).
2. Fitness calculation is carried out in two steps:
   a. Decode chromosomes into an integer, which is called phenotype
      01101-->13, 11000-->24, 01000-->8, 10011-->19
   b. Evaluate fitness value by using function given in the problem statement
      $f(x) = x^2$
      13-->169, 24-->576, 8-->64, 19-->361
3. Selection operation: Select two individuals (or parents) based on their fitness in Pi, which is given as:

$$P_i = F_i / \left(\sum_{J-1}^{n} F_j \right) \quad (6.5)$$

where
F_i = Fitness for string i in population, expand as f(x)
P_i = Probability of string i being selected
n = Number of individuals in population
n * P_i = Expected count (Table 6.1)

As in the above table, string 2 has greater chances to be selected. So, string number 2 is selected for mating to get the offspring with the higher frequency value.

TABLE 6.1 Selection operation

STRING NO.	INITIAL POPULATION	X VALUE	FITNESS F_J $F(X) = X^2$	P_I	EXPECTED COUNT N * P_I
1.	01101	13	169	0.14	0.56
2.	11000	24	576	0.49	1.97
3.	01000	8	64	0.06	0.22
4.	10011	19	361	0.31	1.23
Sum			1170		

4. Crossover operation

It can be of either one point in which there will be one breakpoint and selected pair of strings is cut at some random position and then segments are swapped to form a new pair of strings or two points, in which there will be two breakpoints involved.

Here, the one-point crossover is used. We replace the string that has the least value of the expected count with the string which has the highest value of this count.

If we compare the sum of fitness in the below table with the above table, it has increased from 1170 to 1754. So this is done by genetic algorithm.

5. Mutation operation

It is applied to each child individually after crossover. In this, bits are changed from 0 to 1 or from 1 to 0 at the randomly chosen position of randomly selected strings.

We have not done anything with two and three strings (can do mutation on these two strings) because these strings have the fitness value (see Table 6.2).

TABLE 6.2 Crossover operation

STRING NO.	MATING POOL	CROSSOVER POINT	OFFSPRING AFTER CROSSOVER	X VALUE	FITNESS $F(X) = X^2$
1.	0 1 1 0 1	4	0 1 1 0 0	12	144
2.	1 1 0 0 0	4	1 1 0 0 1	25	625
2.	1 1 0 0 0	2	1 1 0 1 1	27	729
4.	1 0 0 1 1	2	1 0 0 0 0	16	256
Sum					1754

So if we compare the fitness sum after applying the mutation operation (Table 6.3), its value is 2354 from 1754 (Table 6.2). Since GA improved the summation of all the individual fitnesse, we can say that we have better results after applying GA.

TABLE 6.3 Mutation operation

STRING NO.	OFFSPRING AFTER CROSSOVER	OFFSPRING AFTER MUTATION	X VALUE	FITNESS $F(X) = X^2$
1.	_0 1 1 0 0	1 1 1 0 0	26	676
2.	1 1 0 0 1	1 1 0 0 1	25	625
2.	1 1 0 1 1	1 1 0 1 1	27	729
4.	1 0 _0 0 0	1 0 1 0 0	18	324
Sum				2354

9. Advantages
   1. It is effortlessly parallelized, easily modified, and adaptable to different problems.
   2. It has massive and extensive solution space searchability.
   3. It uses payoff (objective function) information instead of derivatives.
   4. It uses probabilistic transition rules instead of deterministic rules.
   5. It is adequate for a noisy environment.
   6. It is vigorous w.r.t. local minima/maxima.
   7. Its concept is easy to understand.
   8. It works exhaustively on mixed discrete/continuous problems.
   9. It is stochastic.
   10. It searches from the population of points, not a single point.
   11. It reinforces multi-objective optimization.
   12. It is faster and efficient as compared to the traditional method which uses the brute force method.

10. Limitations
    1. Its implementation is still an art.
    2. It is computationally expensive i.e. time-consuming.
    3. It is not considered the best solution for simple problems where the derivative information is readily available.
    4. It may not converge to an optimal solution if it is not put in the best manner.
    5. The tough task in GA is the selection of various parameters such as the size of the population, crossover rate, mutation rate, selection method, and its strength.

CONCLUSION

Security-based genetic algorithm serves as a vigorous adaptive approach to remedy exploration and optimization issues. It is based on the catechistic searching technique that solves the problem quickly i.e. guaranteed to give the best solution with the optimal solution and security may be obtained. They are more robust than conventional algorithms. GA does not break easily, unlike in older AI systems. The general GA consists of the following components: initial population, calculate fitness function, selection, crossover, mutation, and stopping criteria. The basic operators used in genetic algorithms are encoding, selection, crossover, and mutation. The security-based GA is useful for medical diagnosis, curing diseases, etc.

AUTHOR'S DETAILS

Dr. Neeraj Kumar Rathore, associate professor, Department of Computer Science, Indira Gandhi National Tribal University (IGNTU-A Central University), Amarkantak, M.P., India
Email: neerajrathore37@gmail.com

Dr. Neeraj joined the Department of Computer Science at Indira Gandhi National Tribal University (IGNTU-A Central University), Amarkantak, M.P., India in March 2021 as an associate professor. Before this, he has more than 13 years of experience of teaching in Sri G.S. Institute of Technology & Science, Indore, M.P., India, Jaypee University of Engineering & Technology, Guna (M.P.), Modi Institute of Technology, Rajasthan, etc., and industrial experience in the IT industry at Computer Sciences Corporation (CSC) in the role of a software engineer. He has a Ph.D. in computer science with a specialization in grid computing (2014) and a M.E. in computer engineering (2008) from Thapar University, Punjab, and a B.E. in computer science and engineering (2006) from Rajasthan University, Rajasthan.

His areas of interest include parallel and distributed computing, grid computing, cloud computing, big data, IoT, DBMS, and data structure. He has over 50 publications in international/SCI journals and conferences and books of repute. Under his supervision, seven master's theses and one Ph.D. have been awarded and two are ongoing.

REFERENCES

[1] Neeraj Rathore and Inderveer Chana, "Load balancing and job migration techniques in the grid: a survey of recent trends." *Wireless Personal Communications*, vol-79, no-3, pp. 2089–2125, 2014.

[2] Vishal Sharma, Rajesh Kumar and Neeraj Kumar Rathore, "Topological Broadcasting Using Parameter Sensitivity Based Logical Proximity Graphs in Coordinated Ground-Flying Ad Hoc Networks", *Journal of Wireless Mobile Networks Ubiquitous Computing and Dependable Applications (JoWUA), SCOPUS indexed*, vol-6, no-3, pp. 54–72, September 2015.

[3] Neeraj Kumar Rathore and Inderveer Chana, "A Cognitive Analysis of Load Balancing Technique with Job Migration in Grid Environment", World Congress on Information and Communication Technology (WICT), Mumbai, IEEE proceedings paper, ISBN -978-1-4673-0127-5, pp- 77–82, December 2011.

[4] Neeraj Kumar Rathore and Inderveer Chana, "Variable threshold-based hierarchical load balancing technique in Grid." *Engineering with computers*, vol-31, no-3, pp. 597–615, 2015.

[5] Neeraj Kumar Rathore and Inderveer Chana, "A Sender Initiate Based Hierarchical Load Balancing Technique for Grid Using Variable Threshold Value" in International conference IEEE- ISPC, ISBN- 978-1-4673–6188- 0, pp. 1–6, 26–28 Sept. 2013.

[6] Neeraj Kumar Rathore and Inderveer Chana, "Job migration with fault tolerance based QoS scheduling using hash table functionality in social Grid computing." *Journal of Intelligent & Fuzzy Systems*, vol-27, no-6, pp. 2821–2833, 2014.

[7] Neeraj Kumar Rathore and Inderveer Chana, "Job Migration Policies for Grid Environment", *Wireless Personal Communication*, Springer Publication-New-York (USA), vol-89, no-1, pp. 241–269, IF – 0.979, July–2016.

[8] Neeraj Kumar Rathore and Inderveer Chana, "Report on Hierarchal Load Balancing Technique in Grid Environment", *Journal on Information Technology (JIT)*, vol-2, no-4, ISSN Print: 2277-5110, pp. 21–35, Sep – Nov 2013.

[9] Neeraj Kumar Rathore and Inderveer Chana, "Checkpointing Algorithm in Alchemi.NET", Annual conference of VijnanaParishad of India and National Symposium Recent Development in Applied Mathematics & Information Technology, JUET, Guna, M.P., Dec 2009.

[10] N Rathore, "Dynamic Threshold Based Load Balancing Algorithms", *Wireless Personal Communication*, Springer Publication-New-York (USA), vol-91, no-1, pp. 151–185, Nov 2016.

[11] Neeraj Kumar Rathore and Inderveer Chana, "Comparative Analysis of Checkpointing", PIMR Third National IT conference, IT Enabled Practices and Emerging Management Paradigm book and category is Communication Technologies and Security Issues, pp no.-32–35, Topic No/Name-46, Prestige Management, and Research, Indore, (MP) India, 2008.

[12] Neeraj Kumar Rathore, "Efficient Agent-Based Priority Scheduling and Load Balancing Using Fuzzy Logic In Grid Computing", *Journal on Computer Science (JCOM)*, vol-3, no-3, pp. 11–22, September-November 2015.

[13] Neeraj Kumar Rathore and Inderveer Chana, "Fault Tolerance Algorithm in Alchemi.NET Middleware", National Conference on Education & Research (ConFR10), Third CSI National Conference of CSI Division V, Bhopal Chapter, IEEE Bombay, and MPCST Bhopal, organized by JUIT, India, 6–7th Mar 2010.

[14] Neeraj Kumar Rathore, "Performance of Hybrid Load Balancing Algorithm in Distributed Web Server System", *Wireless Personal Communication*, Springer Publication-New-York (USA), vol-101, no-4, pp. 1233–1246, IF –1.200, 2018.

[15] Neeraj Kumar Rathore, "Ethical Hacking & Security Against Cyber Crime", *Journal on Information Technology (JIT)*, vol-5, no-1, pp. 7–11 December 2015-February 2016.

[16] Neeraj Kumar Rathore and Anuradha Sharma, "Efficient Dynamic Distributed Load Balancing Technique", in Lambert Academic Publication House, Germany, Project ID: 127478, ISBN no-978-3-659-78288-6, 19-Oct-2015.

[17] Neeraj Kumar Rathore, "Efficient Hierarchical Load Balancing Technique based on Grid" in 29th M.P. Young Scientist Congress, Bhopal, M.P., pp. 55, Feb 28, 2014.

[18] Neeraj Kumar Rathore, "Faults in Grid", *International Journal of Software and Computer Science Engineering, MANTECH PUBLICATIONS*, vol-1, no-1, pp. 1–19, 2016.

[19] Neeraj Kumar Rathore, "Map-Reduce Architecture for Grid", *Journal on Software Engineering (JSE)*, vol-10, no-1, pp 21–30, July-September, 2015.

[20] Neeraj Kumar Rathore, "Efficient Load Balancing Algorithm in Grid" in 30th M.P. Young Scientist Congress, Bhopal, M.P., pp-56, Feb 28, 2015.

[21] Rohini Chouhan and Neeraj Kumar Rathore, "Comparison of Load Balancing Technique in Grid", 17th Annual Conference of Gwalior Academy of mathematical science and National symposium on computational mathamatics & Information Technology, JUET, Guna, M.P., 7–9, Dec 2012.

[22] Neeraj Kumar Rathore and Rohini Chohan, "An Enhancement of Gridsim Architecture with Load Balancing" in Scholar's Press, Project id: 4900, ISBN: 978-3-639-76989-0, 23–Oct–2016.

[23] Neeraj Kumar Rathore, "Installation of Alchemi.NET in Computational Grid", *Journal on Computer Science (JCOM)*, vol-4, no-2, pp.1–5, June-August 2016.

[24] Neeraj Kumar Rathore and Pramod Singh, "An Efficient Load Balancing Algorithm in Distributed Networks" Lambert Academic Publication House (LBA), Germany, ISBN: 978-3-659-78892-5, 2016.

[25] Neeraj Kumar Rathore, "GridSim Installation and Implementation Process", *Journal on Cloud Computing (JCC)*, vol-2, no-4, pp.29–40 August- October 2015.

[26] Neelesh Kumar Jain, Neeraj Kumar Rathore, and Amit Mishra, "An Efficient Image Forgery Detection Using Biorthogonal Wavelet Transform and Improved Relevance Vector Machine with Some Attacks ", *Interciencia Journal*, vol-42, no-11, pp. 95–120, 2017.

[27] Neeraj Kumar Rathore and Pramod Kumar Singh, "A Comparative Analysis of Fuzzy based Load Balancing Algorithm", *Journal of Computer Science (JCS)*, vol-5, no-2, pp. 23–33, June-August 2017.

[28] Neeraj Kumar Rathore "Implementing Checkpointing Algorithm in Alchemi. NET", Master of Engineering Thesis in CSE, Thapar University, June 2008.

[29] Neelesh Kumar Jain, Neeraj Kumar Rathore, and Amit Mishra "An Efficient Image Forgery Detection Using Biorthogonal Wavelet Transform and improved Relevance Vector Machine ", *Wireless Personal Communication*, Springer Publication-New-York (USA), vol-101, no-4, pp. 1983- 2008, IF-1.200, 10-May-2018.

[30] Neeraj Kumar Rathore, "Checkpointing: Fault Tolerance Mechanism", *Journal on Cloud Computing (JCC)*, vol-3, no-4, pp. 27–34, August–October 2016.

[31] Neeraj Kumar Rathore and Harikesh Singh, "Analysis of Grid Simulators Architecture", *Journal on Mobile Applications and Technologies (JMT)*, vol-4, no-2, pp. 32–41, July–December 2017.

[32] Neelesh Jain, Neeraj Kumar Rathore, and Amit Mishra, "An Efficient Image Forgery Detection Using Biorthogonal Wavelet Transform and Singular Value Decomposition " in 5th International Conference on Advance Research Applied Science, Environment, Agriculture & Entrepreneurship Development (ARASEAED), Bhopal organized & sponsored by Jan Parishad, JMBVSS & International Council of people at Bhopal (M.P.) India, pp 274–281, held on 04–06 December 2017.

[33] Neeraj Kumar Rathore, "A Review towards: Load balancing Techniques", *Journal on Power Systems Engineering (JPS)*, vol-4, no-4, pp. 47–60, November 16-January 2017.

[34] Neeraj Kumar Rathore "An Efficient Dynamic and Decentralized Load Balancing Technique for Grid", Ph.D. Thesis in CSE, Thapar University, Nov 2014.

7 Role of High-Performance VLSI in the Advancement of Healthcare Systems

Jeetendra Singh
Dept of ECE, NIT SIkkim

Balwant Raj
PUSSGRC Hoshiyarpur

Monirujjaman Khan
North South University, Bangladesh

Contents

7.1	Smart Healthcare Systems	148
7.2	High-Performance VLSI Architecture in a Healthcare Perspective	150
7.3	VLSI Trends in the Innovation of Healthcare Systems	151
7.4	Current Utilization of VLSI in Healthcare Systems	152
7.5	Utilization of Advanced VLSI Systems in Diagnosis and Prevention of Diseases	154

7.6 Conclusion 157
References 157

7.1 SMART HEALTHCARE SYSTEMS

Healthcare utilizes different types of devices that are implantable, tracking devices, non-invasive, wireless devices, diagnostics, and personalized devices. The government initiative of Digital India has strengthened smart health care. A few innovations include cognitive computing solutions for analyzing diagnostic, clinical, and workflow applications. The new revolution in health care is not only about disease and medicine, but also in using technology to deliver quality and timely care with the help of real-time information that drives safe and efficient patient care. Semiconductor technology can offer to improve healthcare outcomes and reduce costs in healthcare performance. Some examples are significant semiconductor contributions in diagnostic imaging, point-of-care, and distributed healthcare paradigms, etc. VLSI has so many applications in healthcare systems and mainly in biomedical engineering, VLSI devices are grown faster and low-power devices are efficiently used to treat the most of the diseases some of the VLSI applications are implantable medical devices (IMDs), such as defibrillators and pacemakers; medical imaging; neural devices, such as deep brain stimulation (DBS) and prosthesis for the central nervous system (CNS), peripheral nervous system (PNS), cochlear and retinal applications, and ECG and EEG [1,2]. The CMOS7 IC technology is currently widely used in the market. Manufacturing CMOS ASIC8 is more than just bipolar ICs, and merging analogue and digital ICs is a well-known issue. The designer must investigate several methods for effectively implementing analogue functions utilizing CMOS technology. In biological applications like signal processing [3], VLSI technologies such as Xilinx have been employed. In comparison to any other medical imaging method, digital front ends to real-time ultrasound phased array signal processors utilized in ultrasound processing would be more effective. VLSI equates to around 100,000,000 transistors. This includes today's microprocessors, which have over 40–50 million transistors. The chip design is currently at the VLSI level. ULSI (ultra large-scale integration), which some have dubbed [2], is the next level of VLSI, with around 1 billion transistors.

Moreover, the low-cost, lightweight, and portable tele-cardiac [4] system was implemented using FPGA. Deep neural networks (DNNs), artificial intelligence (AI), and virtual augmentation all demand high-speed operations; hence, the hardware implementation of these applications is receiving a lot of attention. However, many parts and complicated interconnections are required, resulting in a huge chip area and high-power consumption. This is accomplished through the use of FPGAs and ASICs. By employing quasi-synchronous implementation, the proposed designs may save up to 33% energy consumption compared to binary radix implementation without sacrificing performance [5].

A multi-objective optimization approach for VLSI circuits was published by Kashfi, F. et al. [6]. Various sorts of power and delay are effectively included in this system. In general, convex models accomplish single-step optimization based on an extra modelling error, but non-convex methods only reach the global optimum if analytical gradient is utilized. Weighted sum, compromise programming (CP), and satisfying trade-off method (STOM) are three approaches that may be used to achieve multi-objective optimization when using an analytical gradient and a convex model. The weight sum method is not recommended for resolving multi-objective optimization problems. STOM is suggested when the designer is interested in a certain topic. V. S. B. Kumar et al. [7] investigated contemporary VLSI design trends, focusing on VLSI architectures with bio-inspired algorithms. The testing findings are based on comparing the levels of performance of several VLSI studies in determining the most exact values. Without the use of bio-inspired algorithms, enhancements such as self-adaptive swarm optimization and VLSI optimum design are proved. The topic of floor planning is measured and analyzed with extreme caution. Cassidy, S. A., et al. [8] gave insight into neuromorphic designs during the nano-CMOS era, which aids in understanding the parallel communication link responsible for the formation of spiking neurons and the spike timing dependent plasticity (STDP) learning circuit. As a result, the neurons are viewed as digital arithmetic logic units and communication processes, opening the way for neural design optimization using spiking neurons and STDP learning, which aids in the validation of design approach with cortical growth potency. Mohana, M. S. et al. [9] used a combination of three distinct processes to compress ECG in remote and zero lossless decompression: strategic execution, Golomb rice coding, and pressing configuration to enlarge the storage room by reducing transmission time. The expectation error was encoded using Golomb rice coding. The pressing configuration was chosen to allow for a continuous interpreting process. The strategic execution is assessed to ensure that more than 48 chronicles for the MIT-BIH arrhythmia data collection are deployed. When compared to the prior lossless ECG compression procedure, this technique produces superior results. A multi-purpose straight indicator setting adaptable Golomb rice achieves a lower sophistication lossless external counter pulsation (ECP) pressure using Xilinx code.

An asynchronous interface, a multi-sensor controller, a register bank, a hardware-shared filter, a lossless compressor, an encryption reader, an error correcting coding (ECC) circuit, a universal asynchronous receiver/transmitter interface, a power management, and a QRS complex detector is included in Chen, S.-L. et al. [10], Chen has preferred VLSI circuit design of the micro control unit for WBSNs. For the output of low-pass, high-pass, and bandpass filters with respect to several body signals, a hardware sharing procedure is used to reduce the silicon area of a hardware-shared filter, where the current encryption coder performs in increasing the average compression rate over 12% in ECG signals, providing body signal analysis and filtering security for WSBNs. A QRS detector has also been created to analyze ECG signals, and an encryption encoder based on the asymmetric cryptography method has been added to protect physical information during data transfer. Chen, A. C., et al. [11] proposed an active VLSI circuit architecture that includes an adaptive fuzzy predictor, voting base system, and tri-stage entropy encoder to improve the proficiency and potency of electroencephalogram (EEG) data transmission across a wireless body area network (WBAN). With the use of the CHB-MIT scalp EEG database, the performance of compression rate was assessed with an average value of 2.35 for 23 channels, and the current approach offered a 14.6% increase in a compression rate and a 37.3% decrease in hardware cost. To improve the performance of the future design, a pipelining approach was applied.

7.2 HIGH-PERFORMANCE VLSI ARCHITECTURE IN A HEALTHCARE PERSPECTIVE

Healthcare, big data, and VLSI technologies are the three basic industries that are widely used in smart healthcare innovations. All the industries are booming with their respective growth, but they have their own limitations. The miniaturization of devices and increase in the measure of number of biomarkers leads to demand for efficient processing and data storage which are achieved through VLSI technologies. The huge amounts of data generated in health care require storing and processing be done effectively. The big data also driving the semiconductor industry is one way. Persistent memory (PM) is silicon-based solid state memory that stores data even in a power failure situation. Persistent memory is very useful in high-performance computing and it is in active development by semiconductor companies. Figure 7.1 shows the utilization of VLSI technology and big data in the advancement of healthcare systems. It includes big data, technology, and health care all together and excludes the other issues that are not related to them.

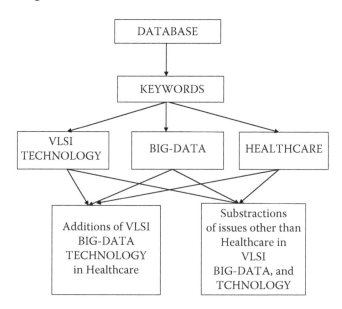

FIGURE 7.1 Utilization of VLSI technology, big data, and health care.

Dielectric modulated device structure [12] also has been developed to recognize the nature of biomolecules which are important to monitor and diagnosis of the diseases. In addition to this novel semiconductor devices nanowire FET, tunnel FET, or nanotube [13–15] can also be employed in this regime to analyze the biomolecules sensitivity which will render fast response and better performance. Memristor is another novel semiconductor device that can be utilized to store the previous record of the patient and their health issues [16]. Its current model is very reliable to realize the past record of the medical science by keeping a large database [17–19].

7.3 VLSI TRENDS IN THE INNOVATION OF HEALTHCARE SYSTEMS

The semiconductor industry growth has been very fast in the past few years and the semiconductor devices are evolving in such a way that we can implement these devices in so many applications and in so many fields like medical, business, construction, manufacturing industries, etc. Electronic system design manufacturing (ESDM) industry is one of the fastest-growing industries in India. Very-large-scale integration (VLSI) is the process that implements the number of ICs in one single chip using hundreds and thousands of transistors and resistors. The VLSI industry has evolved so much in the past few years. And the most of the devices are now used in the medical field also for monitoring the health of the patient and to control the devices that are used for treatment purposes. Also, VLSI devices have been found in automobiles, cell phones, cameras, and a variety of other applications. Advances in process geometry, feature, and product developments have increased the demand for IC design, development, and re-engineering. Core-based design must be investigated to satisfy the growing demand since it has the potential to increase performance.

Also, of great relevance, are network applications. The weight, size, and power of ICs have all improved, making them excellent for medical devices, particularly wearable ones. The importance of VLSI devices in the medical field increased rapidly in the span of less time these devices are used in almost in every equipment that are used in the medical field. Figure 7.2 shows the need of VLSI devices in healthcare systems.

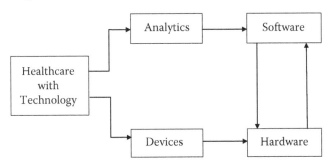

FIGURE 7.2 Implementation of technology in healthcare systems.

Figure 7.2 shows the technology used in healthcare systems. Most of the devices and hardware are designed by using the ICs that are manufactured by using VLSI technology. The device takes the necessary software and gives the output according to the requirements. Most of the devices use VLSI technologies and work accordingly.

7.4 CURRENT UTILIZATION OF VLSI IN HEALTHCARE SYSTEMS

The recent pandemic situation in the healthcare system has disturbed the continuous process of care, leaving the healthcare sector in a "state of emergency" in developing countries like India. The rise of trust flaws between patients and the health bionetwork (due to very poor doctor to patient ratio) might hamper the aspired progress of the Indian healthcare sector and health of the nation. In this situation, Wi-Fi transmission can be accessible with new traditions of healthcare delivery that decreases cost, diminishes victim's distress, and manpower. This wireless structure can be constructed using the very large-scale integration (VLSI) concept, which is appropriate for biogenic applications. VLSI design in biomedicine produces a reduction in size of the chips, range, and speed enhancement. In this report, various proposals are explored for VLSI employment of neural networks, which is stated as a CMOS fabrication technique, architecture of medical implant communication system (MICS) receiver for critical medical operations, field programmable gate array (FPGA) execution of semantic networks, neuro-fuzzy system, neuromorphic computing approach, and neural net performance in analog hardware and digital networks.

Currently, in health care, there exists a big gap among awareness, allocation, and utilization, which leads to a disintegrated system with inadequate access to health care. Though the issues like an enlarged aging trend, accessibility, and privacy occur in the medical care field, the evolution of technologies to solve health problems has developed fast. Now we have the existence of wireless technology, which reduces the patient's pain points, cost, and manpower. Also, there are wireless devices available in the present day that are implanted in the patient's body with the accurate technology for monitoring patients and their medical conditions, the data access is very easy, cost is minimized, and bed space is saved. A multi-objective optimization problem is the streamlining of various functions instantaneously and attaining a solution that is best in respect to all of the objective functions. These problems are present at various levels of VLSI circuit optimization [8].

A medical implant communication system (MICS) is a low-power, short-range (2 m), high data rate (core band is 402–405 MHz) communication network, which has been recognized across the globe for transferring the data to help the diagnostic or analeptic functions linked with medical implant devices [20,21].

Neurology is the medical field concerned with the detection and therapy of ailments of the nervous system, which includes the brain, nerves, and spinal cord. There are more than 600 diseases of the nervous system, which include brain tumors, epilepsy, and Parkinson's disease. Artificial neural networks (ANNs) are computing systems virtually stimulated by the biological neurons that constitute animal brains. Such systems learn to accomplish a task without being automated with particular rules. An ANN uses the processing of the brain as a valid point to build algorithms that can be used to guide complex patterns and prediction problems. An ANN considers data samples rather than the entire data set for any solution that in turn saves money and time. ANNs are networks of computing elements that have the capacity to respond to input stimuli and generate the desired output during VLSI design of neural networks [22,23]. Analog hardware needs to take care with respect to some key aspects: substrate-noise, variations in power supply, drift, leakage, etc. But analog VLSI implementation of ANN by means of a back-propagation algorithm diminishes cost and power dissipation. In order to minimize the power consumption, analog feed-forward neural networks are to be considered for solving the classification problems very easily [24,25]. Digital neural networks are almost produced automatically from a logic description of their functions. Digital ones are well acquainted with new processes and hence redesigning is not required. With these new processes, power and area are lessened in order to make the digital circuits optimized [26,27]. Digital neural networks are highly opted for classification problems even it is with or without analog-digital (AD) conversion of input signals. And moreover, digital networks surpass analog networks when it comes with or without an ADC [28].

The utilization of a field programmable gate array (FPGA) for neural network design gives flexibility in programmable systems. With low precision neural network implementation, FPGAs have faster speed and lesser size for real-time application than VLSI design [29,30]. FPGA plays a very critical role in data sampling and processing industries due to its parallel architecture and low-power consumption [31]. The common neural network construction on the FPGA SOC platform can achieve both forward and backward algorithms in deep neural networks (DNN) with high production and easily gets accommodated to the type and scale of the neural networks [32,33]. FPGAs are hardware accelerators that provide programmable and huge parallel architecture. The mixture of power of GPUs with the reliability of FPGAs extends the scope of problems that can be accelerated [34]. Neuro-fuzzy systems are characterized as special multilayer feed-forward neural networks. This system is educated by a learning algorithm derived from neural network theory. Fuzzy logic makes judgment on the basis of raw and uncertain data given to it. These are used to solve non-linear and complex problems [35,36]. Fuzzy controller executes estimated reasoning on the basis of human ways of perception to gain the control logic. De-fuzzifier alters the fuzzy output to the desired output in order to control the system [37].

In recent times, neuromorphic has been used to discuss about analog, digital, mixed-mode analog/digital VLSI, and software systems that sketches the models of neural systems. The design of neuromorphic computing on the hardware level can be registered by oxide-based memristors, spintronic memories, threshold switches, and transistors [8,38]. Neuromorphic computing systems are highly connected and consume relatively low power and process in memory [39]. Implementation of biomedical systems with the help of VLSI and wireless mechanisms are evolving day by day. Although the analog and digital types have their own advantages, but while coming to design phase, digital ones are considered due to their robustness and ductility [40].

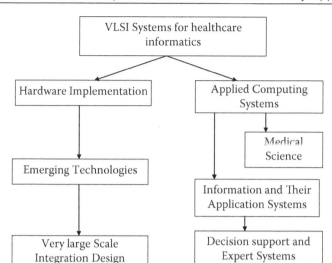

FIGURE 7.3 VLSI systems for healthcare informatics.

Spike sorting is the categorization in which a spike corresponds to a neuron and it is a very challenging problem. With the help of amplitude discriminator, separation of spike with respect to different neuron makes an easy way in terms of fastness and implementation time [41].

Other than these aspects, VLSI also play a vital role in monitoring the vital parameters of patients. Pulse oximeters, digital blood pressure monitors, and blood sugar monitors are a few examples. Figure 7.3 shows the use of VLSI systems in collecting the information of the health issues. The basic need of implementing any hardware relies on the emerging technologies which are developed with the help of VLSI design. In addition, the collected information of any health issue is computed and on the basis of final computed results, a better decision has been made.

7.5 UTILIZATION OF ADVANCED VLSI SYSTEMS IN DIAGNOSIS AND PREVENTION OF DISEASES

The human body is an amazing creation of the universe and it awes scientists and engineers by its amazing operational efficiency in spite of the high complexity. Hence, it becomes quite challenging for healthcare professionals to interpret the signals generated by the body in case of any disease or disorder without the use of any diagnostic equipment. The signals generated in the body are required to be amplified, processed, and displayed in a suitable manner for technicians and medical professionals (doctors and nurses) in order to diagnose and treat a disease/disorder properly.

Hence, electronics have become an indispensable part of medical diagnosis for a very long time. Since heart and brain signals are in the mV and μV range, reliable amplification with better noise immunity is necessary to analyze them. Also, a collection of real-time patient data over a continuous interval of time and early detection of diseases necessitates the development of wearable and implantable electronics. Thus, the focus of VLSI design engineers and researchers is to innovate and provide the healthcare industry with reliable solutions for diagnostic applications.

Discussion: Due to the advances in sensor technologies and imaging modalities such as an electroencephalogram (EEG), intra-cranial electroencephalogram (EEG), magnetoencephalogram (MEG), and magnetic resonance imaging (MRI), healthcare experts can collect data from hundreds of electrodes from the brain at sample rates ranging from 256 Hz to 15 kHz. These data help to analyze brain functioning and brain connectivity at macro and micro levels in healthy subjects to identify patients with neurological and mental disorder. Extracting the appropriate biomarkers can assist clinicians to predict and detect seizures in epileptic patients, and to identify patients with mental disorders such as schizophrenia, depression, and personality disorder. The biomarkers can be tracked to design a personalized therapy and its effectiveness. High-performance VLSI system design is critical to increase battery life of VLSI chips and also for reducing computation time by orders of magnitude in analyzing MRI signals. Monitoring biomarkers and delivering drugs needed are associated with effective analysis of collecting patient data. Extensive research has always been done for developing electronics in the field of health care and it is also an important area at present. Bauer et al. propose an event-driven neuromorphic processor that achieves real-time anomaly detection for a ULP ECG (electrocardiogram) using the concept of a spiking neural network (SNN) [42]. It consumes a total power of 722.1 μW. Ghaemnia et al. present a ULP high-gain CMOS OTA for biomedical applications [43]. It is a post layout simulated using TSMC 180 nm process technology and consumes a total power of 145 nW, including bias circuitry. The double recycling structure, shunt current source, cross-coupled positive feedback, and additional drivers are employed to improve DC gain, GBW, slew rate, and settling response performance without affecting the input referred noise. Danneville uses the spiking neural network (SNN) and presents a novel ULP analog artificial neuron [44]. It uses 65 nm CMOS technology and optimizes power by reducing the supply voltage and related membrane capacitance. Not just speed, but artificial neurons (ANs) improve a lot of aspects of biological neurons (BN). ANs display considerably higher conductivities, much higher velocity (108 m/s as compared to 1–100m/s for BNs), very high spiking frequency (10–100 kHz as compared to 10–100 Hz in BNs), and lower area and lower energy per spike (1fJ as compared to 1pJ in BNs) than their biological counterpart, BN. All the transistors used are operated in deep sub-threshold region of operation. ULP operation is thus achieved by minimizing supply voltage with a power consumption of 100 pW. Orguc et al. present a low-voltage ULP sensor interface for an EMG (electromyogram) application [45]. It mainly consists of an analog front end (AFE) circuitry, an operational transconductance amplifier (OTA), and a SAR-ADC, which consume 3.8 nW, 0.9 nW, and 2 nW pf power, respectively. It can be integrated with sensor nodes as well as implantable devices. Karimi-Bidhendi et al. present two brain signal acquisition (BSA) front-end circuits using ULP CMOS transistors in a weak inversion (WI) regime for an EEG (electroencephalogram) application [46]. Its main focus is to help patients who have suffered a spinal cord injury (SCI) and need a brain computer interface (BCI) to bypass their damaged spinal cord for direct brain control of prosthetics. MOS transistors in the WI region achieve maximum gm/IDS-ratio, resulting in the highest power efficiency at the cost of lower operation maximum bandwidth. The CMRR and power-supply rejection ratios (PSRR) should be large to attenuate the effect of environmental noise sources. Electrical shielding and DC isolation are needed between the IC and implanted electrodes. BSA I incorporates 64 units of amplifier I and a serializer while BSA II incorporates an array of four amplifier II circuits, a serializer, and an instrumentation amplifier. BSAs are fabricated in 130 nm and 180 nm CMOS processes, wherein each amplifier within the arrays consumes 0.216/ 0.69 µW, respectively (not including buffer and in amp). Measured IR noise across the bandwidth was 2.19/2.3 µVRMS corresponding to NEF of 4.65/7.22 and PEF of 11.7/31.3. This research work proves to be a ray of hope for many who had earlier given up on the idea of being able to live normally due to SCI. Luat Tran et al. designed an analog front end (AFE) IC for neural signal acquisition to monitor brain activities [47]. This helps to diagnose and treat neurological disorders such as Parkinson's disease, paralysis, and epilepsy. The AFE consists of a low-noise neural recording amplifier (LNA), a programmable gain amplifier (PGA), an analog buffer, and an ADC driver. It uses inverter stacking topology and results in a power consumption of only 2.8 µW for a 1 V supply. Also, it achieves a low input referred noise of 3.16 µVRMS. Tekeste et al. applies ULP optimization techniques at the architecture and the algorithmic level to design a ULP QRS acquisition and ECG compression architecture for wearables in the domain of IoT used in medical applications [48]. P, QRS, and T are the signal components of an ECG signal used by medical professionals to analyze and diagnose the condition of the heart. The QRS detector consumes only 6.5 nW of power at 1 V supply while the ECG compressor consumes merely 3.9 nW of power at the same 1 V supply voltage. It is implemented in 65 nm low-power process and eliminates the use of multipliers in the circuits replacing them by adders, shifters, etc. Another important aspect of integration of VLSI into medical application is the inclusion of IoT (Internet of Things) sensors in this field. Internet of Things (IoT)–enabled devices have made remote monitoring of patients possible, unleashing the potential to keep patients safe and healthy, and empowering physicians to deliver superlative care. It has also led to an increase in patient engagement and satisfaction as interactions with doctors have become easier and more efficient. Furthermore, remote monitoring of a patient's health helps in reducing the length of hospital stay and prevents re-admissions. IoT also has a major impact on reducing healthcare costs significantly and improving treatment outcomes. Wearable devices can be tuned to remind calorie count, exercise check, appointments, blood pressure variations, and much more. Future works in this domain may help in monitoring chronic patient data and call an ambulance and inform relatives in case of a medical emergency or an early prediction of that using intelligent (AI) systems. Conclusion: Thus, electronics and particularly VLSI systems have proved to be a blessing for patients who are looking for a ray of hope to lead a normal life. It is expected that in the near future VLSI would take huge steps and make early detection of health conditions and consequent treatment easily available to all at reduced, affordable costs, thus, taking humanity towards a new high.

7.6 CONCLUSION

VLSI technology has grown in tremendous ways in the past few years and this technology is used in medical applications to solve so many problems; some of the innovations and the problems are discussed and the importance of the technology is explained. VLSI technology will give more and we discovered that the growth and trends in various industries are intertwined as a result of this report. In the field of health care, technological advancements have exploded. Because of the shrinking of devices and the growth in the number of biomarkers, there is a necessity for efficient processing and data storage, which VLSI technology can provide. On the other hand, the massive volumes of data created in health care must be safely stored and processed. The desire for memory devices that are lower in size and have a higher density necessitates transistor scaling.

REFERENCES

[1] L. Y. Krivonogov, O. N. Bodin, M. N. Kramm and V. G. Polosin. New Technology ECG Signal Interference Suppression. *Journal of Biomedical Engineering and Medical Devices*, 2(127), 2017.

[2] Dr. S. K. Nath and Dr. S. K. Nath. Recent Trends in Medical Imaging by using VLSI Research Inventy: *International Journal of Engineering and Science*, 5(5), pp. 7–10, 2015.

[3] S. Kavitha, M. Arul Pugazhendhi and B. Vinodh Kumar. VLSI Implementation of PPG Signal for Health Monitoring System. *International Journal of Computer Science and Mobile Computing*, 5(3), pp. 667–675, 2016.

[4] Kavya G and Thulasibai V, VLSI Implementation of Telemonitoring System for High Risk Cardiac Patients. *Indian Journal of Science and Technology*, 7(5), 2014.

[5] A. Ardakani, F. Leduc-Primeau, N. Onizawa, T. Hanyu and W. J. Gross. VLSI Implementation of Deep Neural Network Using Integral Stochastic Computing. *IEEE Transactions on Very Large Scale Integration (VLSI) Systems*, 25(10), pp. 2688–2699, 2017.

[6] F. Kashfi, S. Hatami and M. Pedram. *Multi-Objective Optimization Techniques for VLSI Circuits*, University of Southern California, Los Angeles, CA, 2011.

[7] V. S. B. Kumar, P. V. Rao, H. A. Sharath and et al. "Review on VLSI design using optimization and self-adaptive particle swarm optimization," *Journal of King Saud University-Computer and Information Sciences*, 32(10), pp. 1–12, 2018.

[8] S. A. Cassidy, J. Georgiou and A. G. Andreoua. "Design of silicon brains in the nano-CMOS era: Spiking neurons learning synapses and neural architecture optimization," *Neural Networks*, 45, pp. 4–26, 2013.

[9] M. S. Mohana and S. Jayachitra. "An Efficient Code Compression Technique for ECG Signal Application using Xilinx Software," *International Journal of Scientific & Technology Research*, 8(9), pp. 1958–1965, 2019.

[10] S.-L. Chen, C. M. Tuan, Y. H. LEE and et al. "VLSI Implementation of a Cost-Efficient Micro Control Unit with an Asymmetric Encryption for Wireless Body Sensor Networks," *IEEE Access*, 5, pp. 4077–4086, 2017.

[11] A. C. Chen, C. Wu, R. A. P. Abu and et al. "VLSI Implementation of an Efficient Lossless EEG Compression Design for Wireless Body Area Network," *Applied Sciences*, 8(9), pp. 1474, 2018.

[12] G. Wadhwa, P. Kamboj, J. Singh and et al. Design and Investigation of Junctionless DGTFET for Biological Molecule Recognition. *Trans. Electr. Electron. Mater*, 22, pp. 282–289, 2021. doi: 10.1007/s42341-020-00234-8

[13] J. Singh and G. Wadhawa. Novel Linear Graded Binary Metal Alloy PαQ1-α Gate Electrode and Middle N+ Pocket Si0.5Ge0.5 Vertical TFET for High Performance. *Silicon, Springer*, 13, pp. 2137–2144, 2021. doi: 10.1007/s12633-020-00654-4

[14] G. Wadhwa, J. Singh and B. Raj. "Design and Investigation of Doped Triple Metal Double Gate Vertical TFET for Performance Enhancement," *Silicon, Springer*, 13, pp. 1839–1849, 2021. doi: 10.1007/s12633-020-00585-0.

[15] J. Singh, D. Chakraborty and N. Kumar. Design and Parametric Variation Assessment of Dopingless Nanotube Field-Effect Transistor (DL-NT-FET) for High Performance. *Silicon, Springer*, pp. 1–9, 2021. doi: 10.1007/s12633-021-01182-5

[16] J. Singh and B. Raj. "Modeling of Mean Barrier Height Levying Various Image Forces of Metal Insulator Metal Structure to Enhance the Performance of Conductive Filament Based Memristor Model," *IEEE Transaction on Nanotechnology*, 17(2), pp. 268–275, Jan15, 2018.

[17] J. Singh and B. Raj. "Design and Investigation of 7T2M NVSRAM with Enhanced Stability and Temperature Impact on Store/Restore Energy," *IEEE Transaction on VLSI Systems*, 27(6), pp. 1322–1328, 2019.

[18] J. Singh and B. Raj. "Tunnel Current Model of Asymmetric MIM Structure Levying Various Image Forces to Analyze the Characteristics of Filamentary Memristor," *Applied Physics A, Springer*, 125(3), pp. 203–213, 2019.

[19] J. Singh and B. Raj. "An Accurate and Generic Window function for Non-linear Memristor Model" *Computational Electronics, Springer*, 18(2), pp. 640–647, 2019.

[20] R. Venkateswari and R. S. Subha. "Design of MICS Band Low Power Transmitterfor Implantable, Medical Applications," *Advanced Materials Research*, 984, pp. 1223–1228, 2014.

[21] M. N. Islam and M. R. Yuceb. "Review of Medical Implant Communication System (MICS) band and network," *ICT Express*, 2, pp. 188–194, 2016.

[22] A. K. Shrinath. "Analog VLSI Implementation of Neural Network Architecture," *International Journal of Science and Research*, 2, pp. 2319–7064, 2013.

[23] E. Pasero and L. M. Reyneri. "An Analog Cell for VLSI Implementation of Neural Networks," *Neurocomputing*, 68, pp. 153–156, 1990.

[24] L. Y. Song and Vannelli, M. I. Elmasry. "A Compact VLSI Implementation of Neural Networks," *VLSI Artificial Neural Networks Engineering*, 74, pp. 139–156, 1994.

[25] H. P. Graf, L. D. Jackel, R. E. Howard and et al. "VLSI implementation of a neural network memory with several hundreds of neurons," In AIP Conference Proceedings, 151(1), pp. 182, 2008.

[26] M. A. Bañuelos-Saucedo, J. Castillo-Hernández, S. Quintana-Thierry and et al. "Implementation of A Neuron Model using FPGAs," *Journal of Applied Research and Technology*, 1(3), pp. 248–253, 2003.

[27] A. M. Riyaz and B. K. Sujatha. "A Review on Methods, Issues and Challenges in Neuromorphic Engineering," International Conference on Communication and Signal Processing, pp. 2–4, 2015.

[28] S. Venkatesh and P. C. Prasanna Raj. "Analog VLSI Implementation of Novel Hybrid Neural Network Architecture," In Proceedings Analog VI, 7(2), pp. 62–66, 2008.

[29] S. N. Sonar and R. R. Mudholkar. "ECC design strategy targeting area optimizationin reconfigurable device (FPGA) using VLSI technique," *International Journal of Engineering Studies*, 9(1), pp. 1–9, 2017.

[30] S. Murugan, K. P. Lakshmi, J. Sundar and et al. "Design and Implementation of Multilayer Perceptron with On-chip Learning in Virtex-E," In 2nd AASRI Conference on Computational Intelligence and Bioinformatics, pp. 82–88, 2014.
[31] K. Guo, S. Zeng, J. Yu and et al. "A Survey of FPGA Based Neural Network Accelerator," *ACMTransactionson Reconfigurable Technology and Systems*, 9(4), pp. 1–26, 2017.
[32] Y. Qi, B. Zhang, T. M. Taha and et al. "FPGA Design of a Multicore Neuromorphic Processing System," In IEEE National Aerospace and Electronics Conference, pp. 255–258, Dayton, USA, 2014.
[33] F. G. Rodríguez, A. Jiménez-Fernández, F. Pérez-Peña and et al. "ED-Scorbot: A Robotic test-bed Framework for FPGA-based Neuromorphic systems," In 6th IEEE International Conference on Biomedical Robotics and Bio-mechatronics (Bio-Rob), pp. 237–242, Singapore, 2016.
[34] T. Wang, C. Wang, X. Zhou and et al. "An Overview of FPGA Based Deep Learning Accelerators: Challenges and Opportunities," In IEEE 21st International Conference on High Performance Computing and Communications; IEEE 17th International Conference on Smart City; IEEE 5th International Conference on Data Science and Systems (HPCC/SmartCity/DSS), pp. 1674–1681, China, 2019.
[35] S. B. Lakha, R. Goyal, R. Kumar and et al. "Design and VLSI implementation of Fuzzy Logic Controller," *International Journal of Computer and Network Security*, 1(3), pp. 1–5, 2009.
[36] B. M. Wilamowski and R. C. Jaeger. "VLSI Implementation of A Universal Fuzzy Controller," *IEEE Transactionson Industrial Electronics*, 46(6), pp. 1132–1136, 2000.
[37] N. Sadati and H. Mohseni. "A VLSI Neuro-Fuzzy Controller," *Intelligent Automation and Soft Computing*, 5(3), pp. 239–255, 2013.
[38] T. Dalgaty, E. Vianello, B. De Salvo and et al. " Insect-inspired neuromorphic computing," *Insect Science*, 1, pp. 59–66, 2018.
[39] N. K. Upadhyay, H. Jiang, Z. Wang and et al. "Emerging Memory Devices for Neuromorphic Computing," *Neuromorphic Computing*, 4(4), pp. 1–13, 2019.
[40] N. Sowmya and S. S. Rout. "A Review on VLSI Implementation in Biomedical Application," In 10th International Conference on Innovations in Bio-Inspired Computing and Applications, Gunupur, India, 2019. (In Press)
[41] M. Rácz, C. Liber, E. Németh and et al. "Spike detection and sorting with deep learning," *Journal of Neural Engineering*, 17(1), pp. 1–30, 2020.
[42] F. C. Bauer, D. R. Muir and G. Indiveri. "Real-Time Ultra-Low Power ECG Anomaly Detection Using an Event-Driven Neuromorphic Processor," in *IEEE Transactions on Biomedical Circuits and Systems*, 13(6), pp. 1575–1582, Dec. 2019. doi: 10.1109/TBCAS.2019.2953001
[43] A., Ghaemnia and O., Hashemipour. An ultra-low power high gain CMOS OTA for biomedical applications. *Analog Integr Circ Sig Process*, 99, 529–537, 2019. doi: 10.1007/s10470-019-01438-6.
[44] F. Danneville, I. Sourikopoulos, S. Hedayat, C. Loyez, V. Hoël and A. Cappy. "Ultra low power analog design and technology for artificial neurons," 2017 IEEE Bipolar/BiCMOS Circuits and Technology Meeting (BCTM), pp. 1–8, 2017. doi: 10.1109/BCTM.2017.8112899
[45] S. Orguc, H. S. Khurana, H. Lee and A. P. Chandrakasan. "0.3 V ultra-low power sensor interface for EMG," ESSCIRC 2017 -43rd IEEE European Solid State Circuits Conference, pp. 219–222, 2017. doi: 10.1109/ESSCIRC.2017.8094565
[46] A. Karimi-Bidhendi and et al. "CMOS Ultralow Power Brain Signal Acquisition Front-Ends: Design and Human Testing," in *IEEE Transactions on Biomedical Circuits and Systems*, 11(5), pp. 1111–1122, Oct. 2017. doi: 10.1109/TBCAS.2017.2723607

[47] L. Tran and H.-K. Cha. An ultra-low-power neural signal acquisition analog front-end IC, *Microelectronics Journal*, 107, 104950, ISSN 0026-2692, 2021. doi: 10.1016/j.mejo.2020.104950

[48] T. Tekeste, H. Saleh, B. Mohammad and M. Ismail. "Ultra-Low Power QRS Detection and ECG Compression Architecture for IoT Healthcare Devices," in *IEEE Transactions on Circuits and Systems I: Regular Papers*, 66(2), pp. 669–679, Feb. 2019.

Trust-Based Security Model for Adaptive Decision Making in VANETs

Gurjot Kaur
Research Scholar, ECE Department, Dr. B.R. Ambedkar, NIT Jalandha

Deepti Kakkar
Assistant Professor, ECE Department, Dr. B.R. Ambedkar, NIT Jalandhar

Contents

8.1	Introduction	162
	8.1.1 VANET Architecture	163
	8.1.2 Protocols Used for Transmission in VANETs	163
	8.1.3 Security in VANETs	164
	8.1.4 Security Solutions	166
8.2	Gaps in the Existing Solutions	166
8.3	Trust-Based Security Models	167
8.4	Literature on Trust Models in VANETs	168
8.5	Gaps in Existing Trust Model–Based Security Solutions in VANETs	170
8.6	Proposed Trust-Based Adaptive Decision-Making Security Model	171
8.7	Conclusion	174
References		175

8.1 INTRODUCTION

In this modern era, the role of the internet is continuously evolving to facilitate the lifestyle of humankind. Earlier, from being a simple source of communication between a server and a client to hugely revolutionizing modern world, the internet has come a long way [1]. The new advanced technologies collaborating with the internet have turned the cities "smart." In smart cities, the tasks in the cities are handled intelligently without much human intervention through initial integration of cities with information communication technology (ICT) [2]. The global interconnection among the various intelligent objects provide for the extra convenience, comfort, and luxury in today's world. One such exciting research area that has emerged in the context of smart cities is the intercommunication between the vehicles, also termed vehicular ad-hoc networks (VANETs) [3]. VANETs are the ad-hoc networks where vehicles act as the mobile nodes to successfully share information among them [4]. The main idea is to skillfully manage the congested traffic patterns and ensuring the safety of the drivers. Hence, with the integration of dedicated internet resources, enhanced onboard circuitry, and GPS, the vehicles are allowed to participate in an automated network for sharing important information. Mainly, vehicular networks circulate the following three types of information:

1. Emergency messages: Warning messages to be issued to the vehicles in the vicinity whenever an accident happens, potential accident warning whenever a vehicle stops or slows down, assistance messages to avoid any kind of congestion, and emergency warnings in case of any natural calamity like avalanches or landslides [5].
2. Non-emergency commercial messages: The infotainment-related messages like internet services at a particular area, songs, movies, or podcasts of personal choice; downloading a digital map of an area; getting information on available malls, hotels, petrol pumps, etc. are shared [6].
3. Convenience applications: Information related to the electronic toll system for saving fuel [6] and available parking spots are shared.

VANET is a sub-type of mobile ad-hoc networks (MANETs) but also possess other unique characteristics that make their deployment even more challenging. The MANETS are mobile nodes that are not handled by a centralized architecture. VANETs, however, may or may not be handed by a centralized authority. The ease of entering or leaving the mobile nodes in VANETs makes their routing different compared to MANETS [7]. Some of the unique characteristics of VANETs are listed below [8]:

- Highly dynamic network: Due to mobility of the participating nodes, VANET architecture is highly dynamic. There is only a few seconds of link established while communication between two vehicles takes place.
- Discontinuous network: The wireless channels are always uncertain and mobility through these channels having variable attenuation results in frequent disconnection among the vehicles.

- Storage capacity: The VANETs were designed with the aim of providing enough storage capacity and enough battery backup for the vehicles. Hence, there could be long lasting and smoother communications in VANETs as compared to MANETs where the storage and battery resources are limited.
- Mobility modeling: There could be sparse or dense driving environments, different road conditions, different traffic patterns and different behaviors of the drivers in different situations. Accordingly, there are multiple VANET models that need to be addressed individually.
- Interaction using on-board sensors: The easy communication is aided through the on-board circuitry attached to each vehicle that can easily predict the movement of mobile nodes in the vicinity. The detailed architecture of VANETs has been presented in the next section.

8.1.1 VANET Architecture

In the vehicular networks, the vehicles generally communicate directly with each other; vehicle to vehicle (V2V) and with the infrastructure vehicle to infrastructure (V2I) [9]. In [10], it has been found that the network connectivity in VANETs could be improved multi-fold with the help of certain road side units (RSUs) assisting the base stations and the vehicles for message dissemination. The RSUs reduce the rehealing delay, fragmentation problem, and message penetration time. Hence, the VANET architecture mainly comprises the connected vehicles (CVs), RSUs, and on-board units (OBUs) that have the sensory devices and radios, GPS [11] integrated with the proxy, and application and administration servers as shown in Figure 8.1.

The Federal Communications Commission (FCC) in the United States has assigned a dedicated spectrum of 75 MHz specifically for the VANET operation in the 5.9 GHz band called dedicated short range communications (DSRC) [12]. DSRC along with WAVE (wireless access in vehicular environment) and IEEE 802.11p are employed for the design of standard communication protocols for the communication in VANETs.

8.1.2 Protocols Used for Transmission in VANETs

For the data collection and distribution to occur in a systematic way, mainly two types of protocols are being followed in VANETs: unicast and broadcast [13]. The packets are routed to the different mobile nodes through these routing protocols. These are briefly described as follows:

1. Unicast: The data packets are sent to one specific destination by a specific sender through the wireless medium. The packets could be routed through single hop if the sender and receiver are nearby or via multi-hop where the packet is transmitted via hopping to a neighboring vehicle. The VANETs also follow the carry and forward technique where a packet is carried by the most suitable contender (a cluster head) and is transmitted to the destination when needed.

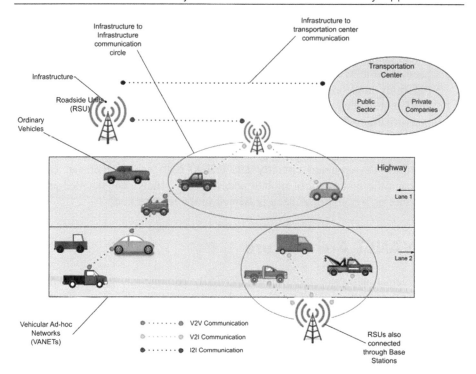

FIGURE 8.1 VANET architecture.

2. Broadcast: These protocols tend to deliver the packets to as many participating CVs as possible. In case of traffic jams, emergencies, or road blocks, broadcasting could be extremely helpful. However, it may lead to congestion in the network because of rebroadcasting.

8.1.3 Security in VANETs

The deployment of VANETs is still an unrealized target due to the primary security concerns involved in the communication. Due to the open access environment and highly dynamic nature of the network, the VANETs are prone to highly complicated attacks. The attacker can easily temper, create, tunnel, inject, or remove the essential information from the network, which could even be fatal for participants. The untrustworthy vehicles perform unreliably in the network and hamper the smooth VANET operation [14]. The broad security attacks are divided into five domains [15,16]. Attacks on: 1. Availability 2. Authenticity 3. Confidentiality 4. Data integrity 5. Repudiation attacks. The various attacks that are covered under these categories are shown in Figure 8.2 and their detailed description is provided as follows:

1. **Availability attacks:** The availability attacks restrict the availability of the resources in the network according to the requirement of the CVs. These are of the following types:

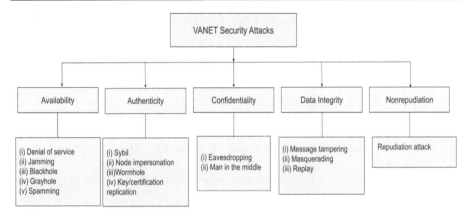

FIGURE 8.2 Security attacks in VANETs.

- Denial of service attack (DOS): The attacker blocks the bandwidth and the available resources by jamming the network [17]. When multiple attackers collaborate to jam the network in a distributed way, it is called the distributed denial of service (DDoS) [18].
- Jamming attack: The attacker blocks the channel by using an equivalent sharp frequency [19] and making it unavailable for the user.
- Blackhole attack: The attacker discards the information that it received and is supposed to communicate further in the network [20].
- Grayhole attack: When the attacker relays some of the information in the network and discards rest of the information [21].
- Spamming attack: When the attacker sends multiple spam messages in the network utilizing a lot of bandwidth [22].
2. **Authenticity attacks:** These attacks aim at procuring the resources without having permission to access them. These are broadly of the following types:
    - Sybil attack: In this type of attack, the attacker possesses multiple fake identities and circulates messages in the network through all of these identities. The receiver may assume it as several other vehicles participating in the congested network [23].
    - Node impersonation attack: When the attacker acquires a legitimate ID of an authorized participant and uses it for another authorized user in the network [24].
    - Wormhole attack: Two or more attackers can place themselves at certain positions in the network and can tunnel the whole conversation in the network to one another by claiming each other as neighbors [17].
    - Key replication attack: An unauthorized node can participate in the network by acquiring duplicate legitimate keys or certificates [25].
3. **Confidentiality attack:** When the content of the messages is modified by the attackers who are not authorized to have access to that information. These are of the following types:
    - Eavesdropping: When the unregistered users listen up to the confidential information not meant for them [26].

- Man in the middle attack: When the attacker tries to get a hold of the private conversation between two vehicles and controls the communication by modifying the contents of the messages [27].
4. **Data integrity attacks:** The intentional attacks modify the content of the messages transmitted in the network.
   - Message tampering attack: When the attacker modifies the content of the information for selfish reasons [28].
   - Masquerade attack: When unregistered attacker enters the network and sends wrong information using a legitimate ID of the registered user [25].
   - Replay attack: When a legitimate message is replayed again in the network at the wrong time [29].
5. **Repudiation attacks:** When an attacker refuses to have participated in either transmitting or forwarding the fraudulent information [30].

8.1.4 Security Solutions

The various solutions that are suggested in the literature for all of the previously mentioned security concerns are briefly shown in Table 8.1 [20,30–33].

8.2 GAPS IN THE EXISTING SOLUTIONS

1. Cryptography-based solutions are robust but they fail in the delay-sensitive scenarios where the message delivery is required to be very fast. The complicated encryption and decryption techniques require a certain time for the receiver to comprehend whether it is the legitimate message or not and whether it could be further transmitted in the network. This causes a certain delay which can be dangerous in the emergency situations where decisions have to be made without any delay.

TABLE 8.1 Solutions of various security attacks in VANETs

Security domain	Best-suited solutions in the literature
Availability	Bit commitment and signature-based authentication
Authentication	Digital signature–based authentication, bit commitment, strong cryptography solutions, zero knowledge–based schemes, keeping check of the legitimate certificates through CRL
Confidentiality	Encryption of data using sophisticated cryptography techniques
Data Integrity	Group key management, digital signatures, zero knowledge techniques
Repudiation attacks	Trusted hardware that prevents any change in protocols, identity-based signature, complex certification management

2. The cryptography and digital signature based solutions involve huge complications and require a dedicated centralized infrastructure for the proper management of the sophisticated keys as well as signatures. As new keys and signatures have to be generated for every single communication, it would be very resource consuming to manage huge amounts of keys, especially in very dense traffic patterns with limited energy resources.
3. The computational complexities are very high in case of cryptography solutions, which is again the waste of resources.
4. Failure of cryptography-based solutions in the case of insider attacks. When an authorized node with all the legitimate keys and certificates performing well in the network tries to misbehave at some time, the VANET gets disrupted in that case. The cryptography-based solutions provide a robust prevention against outsider attacks but fail in the case of insider attacks.

Considering all these gaps associated with the proposed solutions, there existed a need of a simple model that can help in the security of VANET communication without additional complications. A trust-based solution has been discussed in the literature as a simpler alternative to fill the gaps associated with cryptography-based solutions [34]. In trust-based security models, only those messages are accepted that are from the trustworthy node and the rest are discarded. A confidence is put in any vehicle before accepting or forwarding the message that has been transmitted. Any vehicle performing unreliably in the network loses its reputation value and, hence, is not further allowed to participate.

8.3 TRUST-BASED SECURITY MODELS

Broadly, three categories of trust-based models have been discussed in the literature [35–37], as shown in Figure 8.3. These are briefly described as follows:

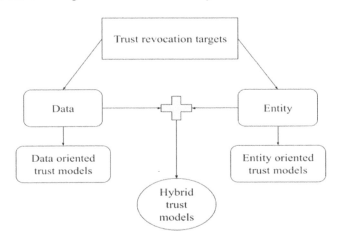

FIGURE 8.3 Three domains of trust-based models.

1. Entity-oriented trust models: These models aim at excluding the malicious vehicles from the network by calculating the trust values of the entities (nodes). These models emphasize initializing the communication based on the reputation of the vehicle.
2. Data-oriented trust models: These models access the trustworthiness of the content of the data. It is hugely dependent on the recommendations from the nearby vehicles that have interacted with this vehicle and can rate its performance based on the past interactions.
3. Hybrid trust models: These types of models combine the properties of both entity-oriented and data-oriented trust models. The decision is made collectively by judging the reputation of the vehicle as well as the recommendations from the vehicles in the vicinity.

The trust-based models are very simple and could perform well in delay-sensitive scenarios as well. They do not, however, provide as much robustness as the cryptography-based techniques, but can provide an additional level of security for the insider attacks when used along with the cryptography-based solutions. Also, when the information to be relayed in the network is not very sensitive, trust models alone can provide good accuracy for the communication. There would not be a need for the cryptography-based encryption for less sensitive information dissemination that can save huge resources in the network. These models do not require a dedicated infrastructure for their management as opposed to the cryptography-based security solutions. The brief comparison of trust-based security solutions with cryptography-based solutions is shown in Table 8.2.

8.4 LITERATURE ON TRUST MODELS IN VANETS

Several trust models have been discussed for the security of VANETs in the literature [38–41]. Trust calculation using direct and indirect interactions called T-V Nets has

TABLE 8.2 Comparison of trust-based security solutions and cryptography-based security solutions

FEATURES	CRYPTOGRAPHY-BASED SECURITY	TRUST-BASED SECURITY
Architecture	Centralized dedicated infrastructure	Distributed or semi-centralized
Prevention from attacker model	Outsider attacks	Outsider as well as insider attacks
Traffic model	Delay tolerant	Delay sensitive and delay tolerant
Accuracy	High	Medium
Computational complexity	High	Low

been proposed in [42]. The calculation is made by confirming the information as forwarded by the beacon messages and recommendations from the one-hop neighbors. The recommendations shared by the neighbors are also forwarded to the RSUs to keep a check on their performance. In [43], each vehicle forms an individual view of the trust values of their neighbors and share it with some trusted authorities. The trusted authorities then form a global view of the recommended trust values and communicate to the CVs in their respective areas. However, this trust model is prone to collusion attacks where many malicious vehicles can collude to bad mouth a genuine node or a malicious node is told as genuine. In [44], direct interactions from the past are used for the trust calculations. Firstly, the authentication of node is performed to check a registered node. Then, the time of the event in message is checked. If it is within the threshold timestamp, the location of the vehicle in the message is confirmed. The trust value is hence determined by employing a fuzzy logic process.

For evaluating the cooperation between vehicular nodes, a game theoretic approach has been designed in [45]. The behavior of the individual nodes and properties of the networks are exploited to attain trust for every crucial VANET application. The game theory approach has also been applied in [46] where the untrustworthy nodes are excluded by considering three parameters: majority opinion about a particular node, betweenness of the node that is determined by the part played by a node in the network, e.g., a node who has been a relay node multiple times, would be trusted by numerous nodes and third parameter is node density, which helps in forming the stable clusters in the network of the CVs with similar speed and direction. In [47], long-term trust values are maintained at the back-end servers and RSUs act as the intermediary nodes between the CVs and the back-end servers. In [48], infrastructure-based trust calculation is employed where every event is reported to the vehicles and the central trust authority. Whenever, a message is received, the trust value of the originator and the forwarders are requested from the central authorities. These types of trust models can put an extra overhead on the central authorities to manage huge amounts of data in the network.

In [49], a different approach for trust calculation is considered. Instead of using entity-oriented or data-oriented trust evaluation, the trust of a particular road segment is evaluated. Multiple plausibility checks are employed to check whether the message origination is from a trusted road segment, whether its integrity or the contents have been compromised during communication through that particular road segment, or whether there could be any other potential attacks. The database of all the road segments and their corresponding trust values are stored. Any entity or vehicle that misbehaves in a particular road segment decreases the reputation value for all the vehicles on that road segment. This technique is particularly unfair to all the innocent participants, whose opinions are completely discarded due to one or more malicious nodes available in that road segment. Moreover, the malicious nodes can keep on misbehaving at different road segments, disrupting the whole network. Various trust models based on the misbehavior detection of the malicious nodes have also been suggested in the literature [50,51]. In [51], both a misbehavior detection technique and trust assessment have been simultaneously employed. The local misbehavior detection scores (MDSs) are used by every CV to calculate their corresponding trust values and share it with the neighbors. In the second approach, only MDSs are shared while trust scores are individually calculated based on the majority of the opinions and consensus.

In the recent years, blockchain technologies have seen a huge development in multiple fields including cybersecurity and security of the computer networks. Various trust models have also been suggested in the literature based on blockchain technologies. In [52], whenever a message is circulated in the network, a CV is selected to rate the accuracy of the message which then stores it in a block and shares it with the neighbors. The neighbors can agree or disagree with the rating based on their knowledge. The majority of the opinions as generated in the network are kept in the block and then added to the blockchain. In another work [53], the actions of the misbehaving nodes and the innocent nodes are checked by a certification authority (CA) through anonymous certificates and are stored in the blockchain. The reputations are then calculated through direct trust, recommended trust values from the neighbors and the misbehaviour score stored in the blockchain. In [54], the CVs assign a rating to a sender based on the bayesian inference model and forwards the rating to their respective RSUs. Further, the trust scores are generated by the RSUs based on these ratings and the final scores are embedded into a block which is then added to the blockchain. The trust values are updated by the competing RSUs through a consensus algorithm.

In [55], a clustering algorithm has been proposed that carefully assigns a role to the participating vehicle as a cluster member or a cluster head. The cluster formation is mainly decided by the similarity in the vehicle mobility. Finally, the establishment and update of trust in the clusters is proposed using machine learning. This algorithm considers both entity- and data-oriented trust models. In [56], the trust values of the vehicles are generated by maximizing path trust values. Based on the assigned trust values, an optimal path for routing the packets in the network is selected using deep reinforcement learning algorithm. In [57], a trust-based model is used for the intrusion detection and support vector machine (SVM) is employed to train the network. There is a significant drawback associated with trust models employing machine learning: they result in even worse bootstrapping problems.

As discussed in the earlier sections, the cryptography-based security solutions are computationally complex and expensive. But for highly sensitive information flow, cryptography-based trust models have also been discussed in the literature [58–60]. In [59], long-term trust is being emphasized in order to save the redundancies of recalculating trust values of the neighbors that were separated for some time but reunited again. The long-term trust is shared by using PKI and certificates. The use of certificates is primarily applied only at the bootstrapping process to save resources. However, the full performance evaluation has not been discussed in this work.

8.5 GAPS IN EXISTING TRUST MODEL-BASED SECURITY SOLUTIONS IN VANETS

1. Initial trust bootstrapping: Many techniques in the literature lack the discussion of assigning the initial trust value to a new vehicle who has not been interacted before. Many techniques prefer to keep the trust value low that

requires a lot of time for the node to gain an appropriate reputation in the network. If it is kept high, the performance of a node is overestimated in the network. In most of the works, the initial trust is designated as neutral and increased and decreased according to the succeeding performances. However, it is not normally the actual trust value of the encountered node. There needs to be a precise way to determine the actual trust values of the newly joined CVs.
2. Coarse-grained revocation in trust models: Whenever a node does not perform reliably, its reputation in the system is decreased. If it is reduced below a certain threshold, the opinion of the corresponding node is discarded and it is not allowed to further participate in the network. However, in realistic scenarios, some packets might be lost due to attenuation, weather conditions, or collisions. Moreover, some participants might have less experience in the use of internet-based applications and their actions might be delayed. But, this may lead to their revocation. So, there is a need to design a method that considers these practical scenarios while making a decision of the reputation value of a node.
3. Lifetime of trust value: The vehicles need to store the updated trust values of their neighbors. However, there could be hundreds of neighbors in dense traffic areas and the storage of a node is limited to store all these values. Hence, a proper lifetime of these trust values for any trust model needs to be explored.
4. Trade-off between accuracy and delay in decisions: For the recommendations-based trust models, there exists a dilemma of number of recommendations that are enough to make an accurate decision. The higher the recommendations, the greater the accuracy but greater delay in making the decision. This trade-off needs to be addressed explicitly while designing the trust models.

8.6 PROPOSED TRUST-BASED ADAPTIVE DECISION-MAKING SECURITY MODEL

VANETs are highly dynamic networks due to their node mobility, but the ever-changing wireless network characteristics can make their deployment even more challenging. The existing trust-based models are rigid in nature that are not suitable to the dynamic nature of the VANETs. Hence, there is a need for the trust models that can adapt to the changing nature of the network and make decisions adaptively for secure communication. The proposed trust model is designed, keeping in view these changing network characteristics as well as introducing subjectivity to prevent the coarse-grained trust revocation in the existing models. Moreover, the emphasis has also been made to the initial trust bootstrapping when a new node joins the communication. This parameter is generally assumed randomly in the existing trust-based models. The layout of the proposed trust-based adaptive security model is shown in Figure 8.4.

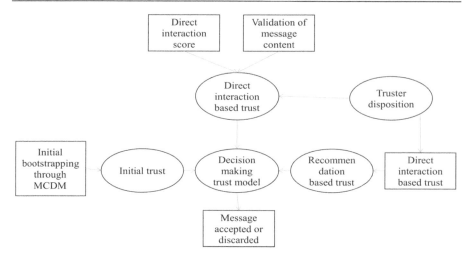

FIGURE 8.4 Layout of the proposed trust-based adaptive decision-making model.

In the proposed model, the final decision of whether to accept or reject a message is made on the basis of overall trust measure evaluated by considering three trust parameters: 1. Initial trust evaluation 2. Direct interaction-based trust 3. Recommendation-based trust.

1. **Evaluation of the initial trust**: In a practical VANET scenario, it is not possible to get a direct trust value and recommendations from other vehicles if there have been no interactions in the past with the newly encountered node. So, the overall trust measure completely depends on the initial trust parameter only. This parameter has to be selected very carefully rather than just assuming a neutral value. The proposed model employs a multiple criteria based decision making (MCDM) process to bootstrap the initial trust values.

Since trust is a subjective term, in this work, subjectivity is being incorporated in order to avoid any sort coarse-grained revocation. The Brown Gibson model [61] is a MCDM technique that adequately makes the decision considering both objectivity and subjectivity of the criteria. The baseline approach is that the reputation R of the vehicle is not a single value but it is a set of values relative to the performance in the vehicles in different context, i.e., $R = (r_{c1}, r_{c2}, \ldots r_{cn})$ where $C = (c_1, c_2, \ldots c_n)$ are the n different contexts. The reputation values are hence derived in multiple contexts e.g. the alteration of message content, generating false information, initiating attacks, or not forwarding the emergency messages on time, result in a very low reputation value. However, the reputation values may not be decreased that sharply when the packets are lost or discarded due to traffic congestion or the commercial and convenience messages are not forwarded on time. This would help to determine the performance of a vehicle in multiple contexts and, based on it, the decisions can be made adaptively in every varying scenario.

The critical parameter in the buildup of initial trust is the past performance of the vehicle. We have measured this parameter as a measure of reliability. Let us assume, whenever a node performs reliably in the network, it is rewarded with a like; similarly, whenever it performs unreliably, it gets a dislike. We calculate the measure of reliability (RM) as:

$$\frac{Number\ of\ Likes}{Number\ of\ Dislikes}$$

The upper limit could be decided for this value e.g. if the likes of the participants are five times that of the dislikes he gets, he may be termed as performing well in the network. So, the upper limit could be assumed based on different applications and then normalized to the range [0–1]. This is the critical objective parameter that we consider in our proposed scheme.

Also, based on the reputation values in different contexts, it can be judged whether the node is deliberately not performing well in the network or there are environmental disturbances or the participant simply lacks the general technical knowledge and his actions are delayed. To accommodate this subjective judgment, a new parameter has been included called experience in the application (EiA). If the user has great experience in the application and understating of the VANET operation but still performs maliciously, he is being given a very low rating in the range [0–1]. However, if he performs unreliably in some situations due to his inexperience, he might receive better EiA score. If he performs well, he gets the highest score of 1.

Now, as the objective and subjective both parameters are discussed, the trust measure could be based on the basis of the Brown Gibson Model. It is mathematically represented as

$$M_i = C_i[W. \ O_i + (1 - W)S_i] \tag{8.1}$$

where M_i is the measure for the ith alternative and C_i is critical factor measure. Its value is either 0 or 1, O is objective factor measure lying between 0 to 1, S is the subjective factor measure between 0 to 1, W is the decision weight of the objective factor. It decides how important the objective parameter is, relative to the subjective parameter for the calculation of the measure M. it could be taken any value. However, the same value would be used by each vehicle for the measure calculation. This measure parameter provides the threshold to decide whether the newly encountered node is to be trusted or not.

2. **Direct interaction-based trust evaluation:** The direct interaction-based trust could be evaluated using past direct interaction score. If there has been a past experience of the interaction with this node, then its prior reputation value is used to determine its performance in the network. The reputation score is then updated through this recent interaction.

For any interaction, time, position, and event details provided in the message are checked by the node. The event details from multiple nodes are compared and a decision is made whether the node is trustworthy or not. Correspondingly, the prior reputation scores at time t are updated as shown:

$$(R_{i,j})^{t+1} = (R_{i,j})^t + \alpha (F_{i,j})^t, \quad \alpha \in [0, 1] \tag{8.2}$$

$$(F_{i,j})^t = f(x^t, y^t, z^t)$$

where $(R_{i,j})^{t+1}$ denotes the updated reputation value at time t+1 when two vehicles i and j interact directly. $(F_{i,j})$ denotes the feedback factor which is determined by the truth value of contents of the message: time (x), location (y), and event similarity (z).

3. **Recommendation-based trust:** This trust value is calculated through the recommendations passed by the nearby vehicles. As every vehicle interacts directly, they form the opinion of the performance of the interacting node and share that trust value with its neighbors. The right recommendations will improve the reputation value of the recommending vehicles by using equation 8.2, where $(F_{i,j})$ would be decided by the truthfulness of the recommendations passed. The false recommendations passed by the nodes would result in degrading their prior reputation values where $(F_{i,j})$ would be a negative quantity. The majority of the recommendations are assumed to be true and the minority opinions are discarded while decision making.

Truster disposition: To account for the subjectivity involved in making decisions in VANETs, a truster disposition parameter is considered in this work. It defines the level of confidence of the participant as he becomes experienced and comfortable with all the technical aspects of the VANET communication. An inexperienced participant is more likely to make mistakes and take longer times to make decisions. This can harm the reputation and trust formation of the vehicle for a very long time in the network. So, every participant can assign itself some value in the range of [0,1] where 0 denotes no experience of the application and 1 denotes good experience. The advantage of communicating this parameter is that an inexperienced person can be allowed some mistakes in the network and is still given a chance to prove himself innocent. This parameter is mainly considered while making the opinion of the vehicle when directly interacting with it. So, even if a small error is performed by an inexperienced vehicle, he might be assigned an average trust score rather than low scores.

Finally, an overall trust score is generated by using the previously mentioned three trust factors, which are then applied to any inference model. The mapping of these trust scores is then done into the low, average, and high trust scores. Based on these scores, the final decision is made on whether to accept a message or reject it.

8.7 CONCLUSION

The vehicular networks face numerous security concerns that still need to be addressed before their deployment. Cryptography techniques do provide the necessary robustness against these security attacks but they put a huge overhead in the system, making it difficult to manage. The simpler alternative to cryptography-based techniques is by employing trust-based models. These models prevent against most of the security attacks in VANETs and do not require any dedicated infrastructure. However, trust models presented in the literature suffer from significant inconsistencies. The major issue that has been considered in this work is the initial trust bootstrapping. The majority of the trust models assign a neutral (0.5) initial trust value to a newly encountered node that is incorrect in most of the cases. So, the proposed work emphasizes resolving this issue by making a decision adaptively for different contexts by keeping in view both the objective and subjective nature of the trust evaluation. Multiple reputation values are considered for different contexts and the final decision is made based on the Brown Gibson multi-criteria decision-making model.

Another contribution of this work includes the prevention of coarse grain revocation of the reputation values due to rigidity of existing trust models. For this, a new parameter, truster disposition, has been introduced, which helps in including the realistic scenarios while making a trust-based decision.

In future work, the proposed model would be implemented and the overall performance would be checked for different VANET scenarios.

REFERENCES

[1] C. C., Aggarwal, N. Ashish, and A. Sheth. "The internet of things: A survey from the data-centric perspective." *Managing and mining sensor data*. Springer, Boston, MA. pp. 383–428, 2013.

[2] B. N., Silva, M. Khan, and K. Han. "Towards sustainable smart cities: A review of trends, architectures, components, and open challenges in smart cities." *Sustainable Cities and Society*, 38, pp. 697–713, 2018.

[3] A., Rasheed, et al. "Vehicular ad hoc network (VANET): A survey, challenges, and applications." *Vehicular Ad-Hoc Networks for Smart Cities*. Springer, Singapore, pp. 39–51, 2017.

[4] R., Di Pietro, et al. "Security in wireless ad-hoc networks–a survey." *Computer Communications*, 51, pp. 1–20, 2014.

[5] H. T., Cheng, H. Shan, and W. Zhuang. "Infotainment and road safety service support in vehicular networking: From a communication perspective." *Mechanical systems and signal processing*, vol. 25, no. 6, pp. 2020–2038, 2011.

[6] V. kumar et al., "Applications of VANET: Present and Future," *Communications and Network*, vol. 5, pp. 12–15, 2013.

[7] P., Ranjan, and K. K. Ahirwar. "Comparative study of vanet and manet routing protocols." *Proc. of the International Conference on Advanced Computing and Communication Technologies (ACCT 2011)*. No. Acct. 2011.

[8] B., Paul, et al. "Vanet routing protocols: Pros and cons." *arXiv preprint arXiv:1204.1201*, 2012.

[9] A. Dua, N. Kumar, and S. Bawa, "A systematic review on routing protocols for vehicular ad hoc networks," *Vehicular Communications*, vol. 1, no. 1, pp. 33–52, 2014.

[10] S. Sou and O. K. Tonguz, "Enhancing VANET Connectivity Through Roadside Units on Highways," in *IEEE Transactions on Vehicular Technology*, vol. 60, no. 8, pp. 3586–3602, Oct. 2011, doi: 10.1109/TVT.2011.2165739

[11] M., Kalinin, et al. "Network security architectures for VANET." *Proceedings of the 10th International Conference on Security of Information and Networks*. 2017.

[12] M. Smita and N. Pathak, "Secured communication in real time VANET," in *Proceedings of the International Conference on Emerging Trends in Engineering and Technology (ICETET)*, pp. 1151–1155, Nagpur, India, 2009.

[13] R., Tomar, M. Prateek, and G. H. Sastry. "Vehicular adhoc network (vanet)-an introduction." *International Journal of Control Theory and Applications*, vol. 9, no. 18, pp. 8883–8888, 2016.

[14] K., Dylykbashi, et al. "Effect of security and trustworthiness for a fuzzy cluster management system in VANETs." *Cognitive systems research*, vol. 55 , pp. 153–163, 2019.

[15] M. Raya and J.-P. Hubaux, "Securing vehicular ad hoc networks," *Journal of Computer Security*, vol. 15, no. 1, pp. 39–68, 2007.

[16] Y. Qian and N. Moayeri, "Design of Secure and Application Oriented VANETs," in *Proceeding IEEE Vehicle Technology Conference (VTC Spring)*, pp. 2794–2799, Calgary, Canada, September 2008.

[17] Zeadally S., Hunt R., Chen Y.-S., Irwin A., and Hassan A. "Vehicular ad hoc networks (VANETS): status, results, and challenges," *Telecommunication Systems*, vol. 50, no. 4, pp. 217–241, 2012.

[18] V. S. Yadav, S. Misra, and M. Afaque, *Security of Wireless and Self-Organizing Networks: Security in Vehicular Ad Hoc Networks*, CRC Press, Boca Raton, FL, USA, 2010.

[19] R. Minhas and M. Tilal, *Effects of Jamming on IEEE 802.11 p systems*, Chalmers University of Technology, Gothenburg, Sweden, 2010.

[20] M. Nidhal, J. Ben-Othman, and M. Hamdi, "Survey on VANET security challenges and possible cryptographic solutions," *Vehicular Communications*, vol. 1, no. 2, pp. 53–66, 2014.

[21] C. A. Kerrache, C. T. Calafate, J. Cano, N. Lagraa, and P. Manzoni, "Trust management for vehicular networks: an adversary-oriented overview," *IEEE Access*, vol. 4, pp. 9293–9307, 2016.

[22] A. Dhamgaye and N. Chavhan, "Survey on security challenges in VANET," *International Journal of Computer Science*, vol. 2, pp. 88–96, 2013.

[23] J. R. Douceur, "The sybil attack," in *Peer-To-Peer Systems*, pp. 251–260, Springer, Berlin, Germany, 2002.

[24] M. S. Al-Kahtani, "Survey on security attacks in vehicular ad hoc networks (VANETs)," in *Proceedings of the Sixth International Conference on Signal Processing and Communication Systems (ICSPCS)*, pp. 1–9, Gold Coast, Australia, December 2012.

[25] M. S., Sheikh, and J. Liang. "A comprehensive survey on VANET security services in traffic management system." *Wireless Communications and Mobile Computing*, vol. 2019, 2019.

[26] M. R., Ghori, et al. "Vehicular ad-hoc network (VANET)." *2018 IEEE international conference on innovative research and development (ICIRD)*. IEEE, 2018.

[27] M. Lal, A. Saxena, V. P. Gulati, and D. B. Phatak, "A novel remote user authentication scheme using bilinear pairings," *Computers & Security*, vol. 25, pp. 184–189, 2006.

[28] A. Rawat, S. Sharma, and R. Sushil, "VANET: security attacks and its possible solutions," *Journal of Information and Operations Management*, vol. 3, no. 1, pp. 301–304, 2012.

[29] B. Parno and A. Perrig, "Challenges in securing vehicular networks," in *Proceedings of the Workshop on Hot Topics in Networks (HotNets-IV)*, pp. 1–6, Maryland, MD, USA, 2005.

[30] M. Azees, L. Jegatha Deborah, and P. Vijayakumar, "Comprehensive survey on security services in vehicular adhoc networks," *IET Intelligent Transport Systems*, vol. 10, no. 6, pp. 379–388, 2016.

[31] H. Hasrouny, A. E. Samhat, C. Bassil, and A. Laouiti, "VANet security challenges and solutions: a survey," *Vehicular Communications*, vol. 7, pp. 7–20, 2017.

[32] M. Muhammad and G. A. Safdar, "Survey on existing authentication issues for cellular-assisted V2X communication," *Vehicular Communications*, vol. 12, pp. 50–65, 2018.

[33] M., Arif, et al. "A survey on security attacks in VANETs: Communication, applications and challenges." *Vehicular Communications*, vol. 19, pp. 100179, 2019.

[34] N., Fan, and Wu C. Q. "On trust models for communication security in vehicular ad-hoc networks.", *Ad Hoc Networks*, 90, pp. 101740, 2019.

[35] J., Zhang. "Trust management for VANETs: challenges, desired properties and future directions." *International Journal of Distributed Systems and Technologies (IJDST)*, vol. 3, no. 1, pp. 48–62, 2012.

[36] J., Zhang. "A survey on trust management for vanets." *2011 IEEE International Conference on Advanced Information Networking and Applications*. IEEE, 2011.

[37] S. S., Tangade, and Manvi S.r S. "A survey on attacks, security and trust management solutions in VANETs." *2013 Fourth international conference on computing, communications and networking technologies (ICCCNT)*. IEEE, 2013.

[38] Z. Movahedi, Z. Hosseini, F. Bayan, and G. Pujolle, "Trust-distortion resistant trust management frameworks on mobile ad hoc networks: A survey," *IEEE Communications Surveys and Tutorials*, vol. 18, no. 2, pp. 1287–1309, 2nd Quart., 2016.

[39] S. Tan, X. Li, and Q. Dong, "A trust management system for securing data plane of ad-hoc networks," *IEEE Transactions on Vehicular Technology*, vol. 65, no. 9, pp. 7579–7592, Sep. 2016.

[40] Y. Wang, I.-R. Chen, J.-H. Cho, A. Swami, and K. S. Chan, "Trust-based service composition and binding with multiple objective optimization in service-oriented mobile ad hoc networks," *IEEE Transactions on Services Computing.*, vol. 10, no. 4, pp. 660–672, Jul. 2017.

[41] Y. Wang et al., "CATrust: Context-aware trust management for serviceoriented ad hoc networks," *IEEE Transactions on Services Computing.*, vol. 11, no. 6, pp. 908–921, Nov. 2018.

[42] C. A. Kerrache, N. Lagraa, C. T. Calafate, J.-C. Cano, and P. Manzoni, "T-VNets: A novel trust architecture for vehicular networks using the standardized messaging services of ETSI ITS," *Computer Communications*, vol. 93, pp. 68–83, Nov. 2016.

[43] C. A. Kerrache, C. T. Calafate, N. Lagraa, J.-C. Cano, and P. Manzoni, "Hierarchical adaptive trust establishment solution for vehicular networks," in *Proc. IEEE 27th Annu. Int. Symp. Pers., Indoor, Mobile Radio Commun. (PIMRC)*, Sep. 2016, pp. 1–6.

[44] S. A. Soleymani et al., "A secure trust model based on fuzzy logic in vehicular ad hoc networks with fog computing," *IEEE Access*, vol. 5, pp. 15619–15629, 2017.

[45] S. Shivshankar and A. Jamalipour, "An evolutionary game theory-based approach to cooperation in VANETs under different network conditions," *IEEE Transactions on Vehicular Technology*, vol. 64, no. 5, pp. 2015–2022, May 2015.

[46] M. M. Mehdi, I. Raza, and S. A. Hussain, "A game theory based trust model for vehicular Ad hoc networks (VANETs)," *Computer Networks*, vol. 121, pp. 152–172, Jul. 2017.

[47] T. Biswas, A. Sanzgiri, and S. Upadhyaya, "Building long term trust in vehicular networks," in *IEEE 83rd Vehicular Technology Conference (VTC Spring)*, May, pp. 1–5, 2016.

[48] C. Liao, J. Chang, I. Lee, and K. K. Venkatasubramanian, "A trust model for vehicular network-based incident reports," in *Proc. IEEE 5th International Symposium* on *Wireless Vehicular Communications(WiVeC)*, pp. 1–5, Jun 2013.

[49] S. Ahmed and K. Tepe, "Misbehaviour detection in vehicular networks using logistic trust," in *Proc. IEEE Wireless Communications and Networking Conference*, pp. 1–6, Apr. 2016.

[50] A. Singh and H. C. S. Fhom, "Restricted usage of anonymous credentials in vehicular ad hoc networks for misbehavior detection," *International Journal of Information Security*, vol. 16, no. 2, pp. 195–211, Apr. 2017.

[51] T. R. V. Krishna, R. P. Barnwal, and S. K. Ghosh, "CAT: Consensusassisted trust estimation of MDS-equipped collaborators in vehicular ad-hoc network," *Vehicular Communications*, vol. 2, no. 3, pp. 150–157, Jul. 2015.

[52] Z. Yang, K. Zheng, K. Yang, and V. C. M. Leung, "A blockchainbased reputation system for data credibility assessment in vehicular networks," in *Proceedings IEEE 28th Annual IEEE International Symposium on Personal, Indoor and Mobile Radio Communications (PIMRC)*, pp. 1–5, Oct. 2017.

[53] Z. Lu, Q. Wang, G. Qu, and Z. Liu, "BARS: A blockchain-based anonymous reputation system for trust management in VANETs," in *Proceedings 17th IEEE International Conference on Trust, Security and Privacy in Computing and Communications and 12th IEEE International Conference on Big Data Science and Engineering(TrustCom/BigDataSE)*, pp. 98–103, Aug. 2018.

[54] Z. Yang, K. Yang, L. Lei, K. Zheng, and V. C. M. Leung, "Blockchainbased decentralized trust management in vehicular networks," *IEEE Internet of Things Journal*. vol. 6, no. 2, pp. 1495–1505, Apr. 2019.

[55] 3. Oubabas, R. Aoudjit, J. J. P. C. Rodrigues, and S. Talbi, "Secure and stable vehicular ad hoc network clustering algorithm based on hybrid mobility similarities and trust management scheme," *Vehicular Communications*, vol. 13, pp. 128–138, Jul. 2018.

[56] D. Zhang, F. R. Yu, and R. Yang, "A machine learning approach for software-defined vehicular ad hoc networks with trust management," in *Proceedings IEEE Global Communications Conference (GLOBECOM)*, pp. 1–6, Dec 2018.

[57] E. A. Shams, A. Rizaner, and A. H. Ulusoy, "Trust aware support vector machine intrusion detection and prevention system in vehicular ad hoc networks," *Computers & Security*, vol. 78, pp. 245–254, Sep. 2018.

[58] T. N. D. Pham and C. K. Yeo, "Adaptive trust and privacy management framework for vehicular networks," *Vehicular Communications*, vol. 13, pp. 1–12, Jul. 2018.

[59] T. Biswas, A. Sanzgiri, and S. Upadhyaya, "Building long term trust in vehicular networks," in *Proceedings IEEE 83rd Vehicular Technology Conference (VTC Spring)*, pp. 1–5, May. 2016.

[60] J. Son, D. Kim, H. Oh, D. Ha, and W. Lee, "Toward VANET Utopia: A new privacy preserving trustworthiness management scheme for VANET," in *Proc. IEEE Int. Conferences Big Data Cloud Computing (BDCloud), Social Computing and Networking (SocialCom), Sustainable Computing and Communications(SustainCom) (BDCloud-SocialCom-SustainCom)*, pp. 301–308, Oct. 2016.

[61] N. Yimen, T. Tchotang, A. Kanmogne, Y. Adamu, F. L. Fon, and M. Dagbasi, "Brown–Gibson Model as a Multi-criteria Decision Analysis (MCDA) Method: Theoretical and Mathematical Formulations, Literature Review, and Applications," *Multiple Criteria Decision Making*, pp. 169–191, 2022.

[62] N., Mani, M., Moh, & T. S., Moh, Defending deep learning models against adversarial attacks. *International Journal of Software Science and Computational Intelligence (IJSSCI)*, vol. 13, no. 1, pp. 72–89, 2021.

[63] J., Sarivougioukas, & A., Vagelatos, Modeling deep learning neural networks with denotational mathematics in UbiHealth environment. *International Journal of Software Science and Computational Intelligence (IJSSCI)*, vol. 12, no. 3, pp. 14–27, 2020.

[64] A., Tewari, et al., Secure Timestamp-Based Mutual Authentication Protocol for IoT Devices Using RFID Tags. *International Journal on Semantic Web and Information Systems (IJSWIS)*, vol. 16, no. 3, pp. 20–34, 2020.

[65] R. K., Dwivedi, R., Kumar, & Buyya, R., Healthcare monitoring sensor cloud with attribute-based elliptical curve cryptography. *International Journal of Cloud Applications and Computing (IJCAC)*, vol. 11, no. 3, pp. 1–18, 2021.
[66] B. B., Mamta Gupta, K. C., Li, V. C., Leung, K. E., Psannis, & S., Yamaguchi, Blockchain-assisted secure fine-grained searchable encryption for a cloud-based healthcare cyber-physical system. *IEEE/CAA Journal of Automatica Sinica*, vol. 8, pp. 1877–1890, 2021.
[67] M., Sarrab, & F., Alshohoumi, Assisted-fog-based framework for iot-based healthcare data preservation. *International Journal of Cloud Applications and Computing (IJCAC)*, vol. 11, no. 2, pp. 1–16, 2021.

Security Attacks and Challenges of VANETs

9

Manojkumar B. Kokare and Deepti Kakkar
*Department of Electronics and Communication Engineering
Dr. B. R. Ambedkar National Institute of Technology,
Jalandhar, India*

Contents

9.1	Introduction	183
9.2	The Communication Topology of VANET	184
	9.2.1 Vehicle to Vehicle Communication (V2V)	184
	9.2.2 Vehicle to Infrastructure	185
	9.2.3 Hybrid Architecture	185
9.3	VANET Characteristics	186
	9.3.1 Mobility Modelling	186
	9.3.2 Frequently Disconnected Network	186
	9.3.3 Network Size	187
	9.3.4 Wireless Communication	187
	9.3.5 Time Critical	187
	9.3.6 Sufficient Energy	187
	9.3.7 Better Physical Protection	187
	9.3.8 Access of Infrastructure	187
9.4	Requirement of Security Services in VANET	187
	9.4.1 Attacks on Integrity	188
	9.4.2 Attacks on Availability	188
	9.4.3 Attacks on Authentication/Identification	188
	9.4.4 Non-Repudiation	188
	9.4.5 Confidentiality	188

DOI: 10.1201/9781003189633-9

9.5	Attacks in VANETS		189
	9.5.1 Attacks on Authentication		189
		9.5.1.1 GPS Spoofing Attack	189
		9.5.1.2 Free Riding Attack	190
		9.5.1.3 Tunneling Attack	190
		9.5.1.4 Replication Attack	191
		9.5.1.5 Message Tampering Attack	191
		9.5.1.6 Sybil Attack	191
		9.5.1.7 Impersonation Attack	191
		9.5.1.8 Wormhole Attack	191
	9.5.2 Attacks on Availability		191
		9.5.2.1 Jamming Attack	191
		9.5.2.2 Black Hole Attack	192
		9.5.2.3 Greedy Behavior Attack	192
		9.5.2.4 Denial of Service Attack	192
		9.5.2.5 Grey Hole Attack	192
		9.5.2.6 Broadcast Tampering	193
		9.5.2.7 Spamming	193
		9.5.2.8 Malware Attack	193
	9.5.3 Attacks on Confidentiality		194
		9.5.3.1 Eavesdropping	194
		9.5.3.2 Man in the Middle Attack	194
		9.5.3.3 Traffic Analysis Attack	195
		9.5.3.4 Social Attack	195
	9.5.4 Attacks on Data Integrity		195
		9.5.4.1 Illusion Attack	195
		9.5.4.2 Masquerading Attack	195
		9.5.4.3 Replay Attack	196
	9.5.5 Attacks on Non-Repudiation		196
		9.5.5.1 Repudiation Attack	196
9.6	Challenges and Future Perspectives		196
	9.6.1 Network Management		196
	9.6.2 Congestion and Collision Control		197
	9.6.3 Environmental Impact		197
	9.6.4 MAC Design		197
	9.6.5 Security		197
	9.6.6 Data Consistency Liability		197
	9.6.7 Low Tolerance for Error		197
	9.6.8 Key Distribution		198
	9.6.9 Incentives		198
	9.6.10 Highly Heterogeneous Vehicular Networks		198
	9.6.11 Data Management and Storage		198
	9.6.12 Localization Systems		198
	9.6.13 Disruptive Tolerant Communications		199
	9.6.14 Tracking a Target		199

9.7 Conclusion   199
Abbreviations   199
References   200

9.1 INTRODUCTION

Technology based on sensors has become more prevalent in the automation industry. Applications can involve many kinds of sensor network–based infrastructure in health care, education, transportation, and more. Wireless sensor networks and mobile ad-hoc networks, which are using in mobile sensing, computation, and communication, are among the domains of VANET. There are three types of mobile nodes in VANETs: sensors on vehicles, statically infrastructure, and roadside units (RSUs) that are connected by wireless connections with vehicle to vehicle, vehicle to infrastructure, and infrastructure to a vehicle. In VANETs, the global positioning system (GPS) is used, as well as cellular communication, either in multiple hop modes, sometimes it happens in single mode also, depending on the desired coverage [1–28]. On this type of technology is primarily designed to ensure the safety of the drivers, leading to fewer road accidents. Providing safety to passengers onboard is one of the important services that this type of network provides [29]. In a VANET, there should be adequate storage, strong power for processing, enough energy, and the ability to estimate the movement of nodes.

There are two types of VANET applications, out of which one is regarded as a critical application, which prioritizes human existence. In addition to diversion notification, emergency response, road conditions information, and prioritizing emergency vehicles like ambulances and fire trucks, it also covers lane handling, congestion minimization, and congestion avoidance. Second, there are numerous add-ons such as toll barrier payment and navigation services.

VANET has a major concern towards the security as, such transportation systems handle the core domain of an intelligent system [30]. In such cases, the security breach directly affects and hinders its performance. There can be many challenges relating to the security of VANETs, concerning new and emerging threats like spoofing in the middle, zero-day attack, etc.; the networks of such types are heavily sensitive to data privacy and to the type of information exchange [31]. Thereby, a proper attention towards any application before its final implementation is necessary. For instance, a piece of information is sent by a naïve moving node to the neighboring mobile nodes and also to RSU and if in case any malicious node discards or tries to alter the information and sends further. As such, either a hazardous or heavily trafficked route can then be modified to avoid the consequences. As a result, such conditions can prove to be highly dangerous, causing some severe and serious injuries or death to the naïve moving node. Thus, in order to avoid such malicious practices, a proper secure system in VANETs is necessary.

The attacker may be able to take advantage of information about the speed, status, and surroundings of the vehicle in a VANET. This includes to the need to take the required steps to avoid such attacks. There can be a misuse of the information of the driver by the attacker. If the driver is constantly thinking about such a situation where his personal information can be misused, then it may directly hit the driver's confidence. The driver's refusal to participate in the network can further result to a black hole problem [1]. We can relate to another problem connected to the metropolitan city with heavy traffic can comprises of a possibility like message flooding for verification. This points to a need of designing a proper routing protocol to handle and cope up with such a serious issue. Also, in addition, scalability and management-related problems in a metropolitan city increase with a large-scale VANET.

The primary thing that relates to the security purpose in VANETs consists of the maintenance of accountability of the vehicle, which is the origin of data. The responsibility of generating the respective messages lie solely to this particular vehicle. The parameters that are of concern to maintain the accountability are authentication, nonrepudiation, availability, and integrity. In the following sections the role of each parameter is talked about. The confidential aspect of the transmitted data is related to the privacy in VANET. Just by continuously watching an attacker on the VANET communication monitoring system may get the security key. This promotes the hidden secrecy from the vehicles those pretend to be truthful and in reality the scenario is different.

9.2 THE COMMUNICATION TOPOLOGY OF VANET

The VANET architecture is categorized into three different kinds that are described as follows:

- vehicle to vehicle communication (V2V)
- Vehicle to Infrastructure (V2I)
- hybrid architectures

All three are divided in accordance with their communication platform.

9.2.1 Vehicle to Vehicle Communication (V2V)

In this type of architecture allows communication between vehicles without being dependent on a fixed infrastructure. This type of communication makes it easier to deal with emergencies as well as gives more safety in communication. Figure 9.1 shows communication between vehicles.

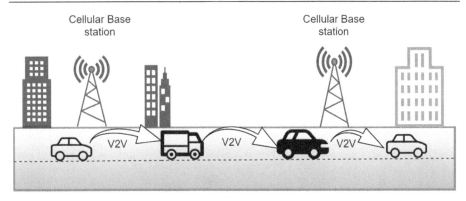

FIGURE 9.1 Communication between vehicles.

9.2.2 Vehicle to Infrastructure

V2I is used for communication between the roadside unit and other infrastructure available within the vehicle trajectory. V2Is are typically used to collect data and information necessary for users. Figure 9.2 shows communication between vehicles and infrastructure.

9.2.3 Hybrid Architecture

This contains both qualities of V2V and V2I as the name also suggests. This type of communication occurs roadside, infrastructure units, and vehicle communication. That

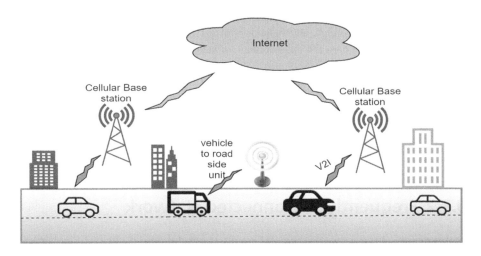

FIGURE 9.2 Communications between vehicles and infrastructure.

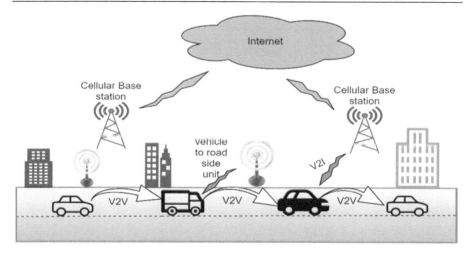

FIGURE 9.3 Hybrid architecture.

communication mode is used for multi-hop as well as long mode of communication. Figure 9.3 shows communication by hybrid architecture.

9.3 VANET CHARACTERISTICS

9.3.1 Mobility Modelling

VANET is describing the mobility pattern that is supporting the surroundings of traffic, movement speed of vehicles, and behaviors of driving style.

Due to dependence on the environment, attackers are attacked easily, and that are defined in the attacks of

- Confidentiality
- Integrity
- Availability
- Authentication/Identification
- Privacy

9.3.2 Frequently Disconnected Network

In this rapidly changing world, vehicles are also faster because of the rapid movement, sometimes information sharing and communication are very important issues. As a result of fast movements, connections are lost and disconnected easily.

9.3.3 Network Size

VANET can be deployed for a city, countries, and worldwide. This means that the VANET network is not bound by geographical boundaries.

9.3.4 Wireless Communication

Designing the VANET is for withstanding a loose wire environment. As a result of these connecting nodes, communication is wireless information. For this reason, certain safety measures must be taken into account when communicating.

9.3.5 Time Critical

In order for a node to act timely, the VAVETs information must be transmitted within a specific time frame.

9.3.6 Sufficient Energy

Nodes of VANET do not have efficient energy for computing resource problems. This makes it possible for VANET to use complex techniques such as ECDSA; RSA is implemented and that delivers endless power.

9.3.7 Better Physical Protection

VANET nodes have improved physical protection. This makes VANET nodes harder to physically compromise and reduces the impact of attack from infrastructure.

9.3.8 Access of Infrastructure

Car mobility makes it very difficult to maintain connectivity. As a result, communication infrastructures such as road units, access points, and hot spots play an essential role in shortening the connectivity time.

9.4 REQUIREMENT OF SECURITY SERVICES IN VANET

When we are talking about any type, the first thing that comes to mind that is security. That plays an important role in any type of communication. An ad-hoc network is an important issue for applications in security-sensitive networks. Then a question comes to mind. How do you secure ad-hoc networks? Then we consider some important criteria for measuring security qualities that contain availability, authentication, non-repudiation, confidentiality, and integrity.

9.4.1 Attacks on Integrity

In this type of attacking, an attacker is adding delay and that creates a problem for users to receive the messages that are coming before the time and without altering the content of messages. That's also known as a timing attack. Because of that users are facing traffic congestion or, worse, accidents. Realizing the above problem, the outcomes that occur from the information and the message delivery in VANET, it is important to be received by users in a given specific time. That is further divided in two types: basic level and extended level. Both are containing the levels in the vehicular network for users. The basic level is used for communicating to the user by the help of P2P and extended level focusing, the things of the basic level when working for a group of users.

9.4.2 Attacks on Availability

In networks, attackers try to different types of attacks based on the availability of resources. The intrusive attack often faced is the denial of service (DOS) against the availability. This denial of service may cause the legal vehicles to lose access to the network by controlling the resources of the vehicle and channels of communication are jammed.

9.4.3 Attacks on Authentication/Identification

In this type of network, attackers enter the network first and then perform the attacks with on-board network unit. In order to carry out any type of attack, an attacker can conceal himself as a legitimate node. One of the easiest attacks that can be carried out against a network is the masquerading attack.

9.4.4 Non-Repudiation

Messages such as accident messages cannot be denied ever being sent or received by either party. The ability to prove RSUs and vehicles have been received and sent is called audit ability in certain fields.

9.4.5 Confidentiality

The network adheres to confidentiality, which means classified information cannot be divulged. A geolocation service can protect sensitive information like names, plate numbers, and locations from unauthorized access as well. When it comes to vehicular networks, pseudonyms are the most popular privacy preservation method. A pseudo, which is used to encrypt or sign messages, is not linked to the vehicle node, but it is accessible to relevant authorities. A new pseudo must be obtained by the vehicle from its RSUs before the old one expires.

9.5 ATTACKS IN VANETS

VANET communication is attacked by various kind of attacks. Generally, the attacker tries to mislead the whole VANET communication, delay the service, modify the message, jam the message, and change the format of communication. The attacks are divided in different types, shown in Figure 9.4.

9.5.1 Attacks on Authentication

9.5.1.1 GPS Spoofing Attack

GPS plays an important role in the VANET network, which gives the exact position of vehicle. When GPS spoofing occurs, an antenna is used to send a counterfeit GPS signal to counter a legitimate satellite signal from a GPS transmitter by attacking the GPS system. The attackers manipulate the information regarding position and time by

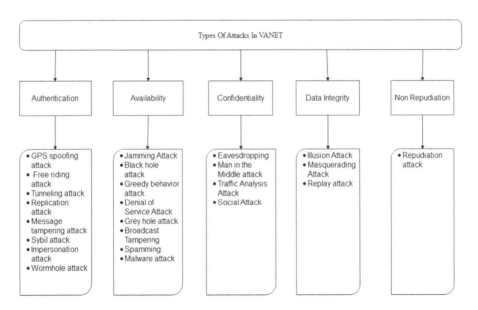

FIGURE 9.4 Types of attacks in VANET.

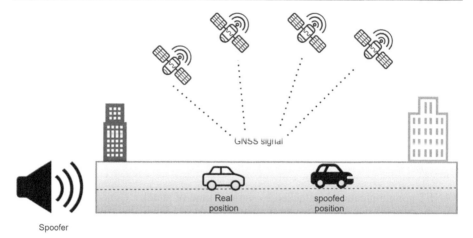

FIGURE 9.5 GPS spoofing attack.

overpowering the GPS signal and the receiver is spoofed with a fake GPS signal. GPS spoofing is shown in Figure 9.5.

9.5.1.2 Free Riding Attack

In this attack the attacker is a free rider. The vehicle node tries to make use of other nodes in the network but it does not return to the network. It tunnels the TCP header and makes false communication.

9.5.1.3 Tunneling Attack

In this type of attack, the attacker places two nodes anywhere in a network using an external communication channel that acts as the neighbor and shares data. A tunnel causes a problem by sharing data between the other nodes in the network and the attacker has access to all of the node data. Figure 9.6 shows a tunneling attack.

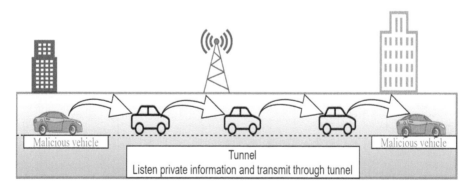

FIGURE 9.6 Tunneling attack.

9.5.1.4 Replication Attack

In a replication attack, the attacker tries to add malicious nodes in the network that uses the identity of another node in the network and transmits the fake data in the network.

9.5.1.5 Message Tampering Attack

The message sent by the transmitter is sent through the wireless medium in a VANET network so it is so much easier to tamper with the message. The original message is first received by a malicious node and it tampers with the message and sends it to the receiver node. So, the unification of the message must be safe the entire time.

9.5.1.6 Sybil Attack

In a sybil attack, the attackers send fake messages to the entire VANET network and control the network, which causes confusion in the VANET network and communication between vehicles nodes is compromised, which eventually reduces the consistency and efficiency of the VANET network.

9.5.1.7 Impersonation Attack

In this attack, the fake node is implemented by the attacker who is impersonates the original node and all other nodes consider that node as a trustworthy node.

9.5.1.8 Wormhole Attack

An attacker in a VANET can control two VANET nodes through a wormhole attack, which tunnels packets between two malicious nodes. This wormhole attack can be classified into three types: open, half open, and closed.

i. Open: Nodes that are malicious consider themselves to be a part of an infected request packet header and assume that the neighboring node is malicious.
ii. Half open: The messages can only be switched from one side to another.
iii. Closed: There is a tunnel connecting the two sides of the wormhole.

9.5.2 Attacks on Availability

9.5.2.1 Jamming Attack

In this attack, attackers try to decrease the signal to noise ratio of the communication network by sending a constant radio frequency signal. In this attack, the attacker is mainly jamming the MAC layer (Figure 9.7).

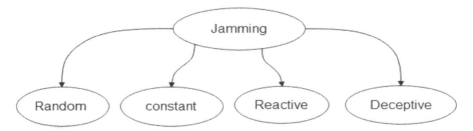

FIGURE 9.7 Types of jamming.

The jamming attack is classified into four types according to their behavior:

i. Random jamming: In random jamming, the attacker saves energy in both sleep and jam interval. In sleep interval, it acts silent but in jam interval it acts as a dominant jammer.
ii. Constant jamming: Attacker sends constant rate data regardless of whether the channel is idle or not.
iii. Reactive jamming: If the channel is in idle mode, it remains in inactive mode but as it senses the channel is transmitting it activates and tries to drop the packets used for transmission.
iv. Deceptive jamming: In this jamming communication, overhead is created by transmitting the constant packets without giving any time gaps.

9.5.2.2 Black Hole Attack

In this attack, the attacker presents a malicious node as an authentic node and attacks the VANETs routing protocol and gives false information to the user nodes and fools them.

9.5.2.3 Greedy Behavior Attack

A greedy behavior attack is done by an attacker with greedy intent. An attacker tries to use the VANET network for himself so he can divert all other vehicles to the different route to get the route cleared and traffic free path for themself.

9.5.2.4 Denial of Service Attack

In a denial of service attack, the communication between vehicles denies the attacker. It causes the packet drop in the communication and also delay the packets and the user is unable to perform necessary tasks. This attack is widely experienced in VANET. Figure 9.8 shows a denial of service attack.

9.5.2.5 Grey Hole Attack

This attack mainly targets the network layer in a VANET network. A gray hole attack sends packets partially or drops the remaining packet in the transmission. It causes the network to overload and reduces the packet delivery rate.

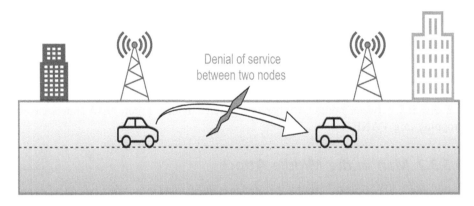

FIGURE 9.8 Denial of service attack.

9.5.2.6 Broadcast Tampering

In VANET communication, every message is considered a broadcast message. In broadcast tampering, communication messages are modified and also packets are dropped and denied.

9.5.2.7 Spamming

In spamming, the unwanted or spam messages are sent by attackers in the form of a message and it causes the delay in communication.

9.5.2.8 Malware Attack

The malware or virus is inserted by attackers in the communication channel, and this virus causes problems in communication. Figure 9.9 shows a malware attack.

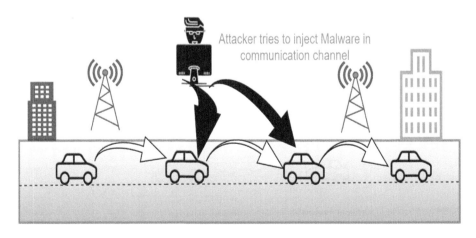

FIGURE 9.9 Malware attack.

9.5.3 Attacks on Confidentiality

9.5.3.1 Eavesdropping

The conversation between two nodes is secretly listened to by another unauthenticated node, called eavesdropping. Such nodes share the communication parameters and conversation between nodes to other unregistered nodes. Figure 9.10 shows eavesdropping in a VANET.

9.5.3.2 Man in the Middle Attack

In this attack, the message sent by the sender is interrupted by a malicious attacker and the modified message is sent to the receiver. Because this happens, the receiver receives the wrong information from attackers, but he assumes that this is the true and authentic information directly from the sender. Figure 9.11 shows a man in the middle attack in a VANET.

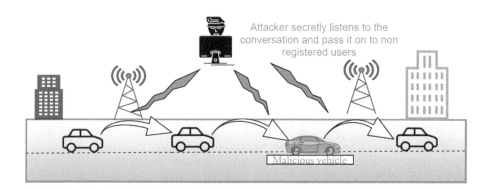

FIGURE 9.10 Eavesdropping.

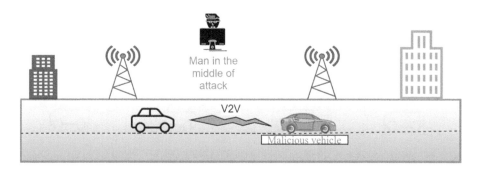

FIGURE 9.11 Man in the middle attack.

9.5.3.3 Traffic Analysis Attack

In such an attack, the attacker gathers all the information about the traffic flow in the communication network and it analyzes the communication frequency and then attacks the network to do some malicious acts.

9.5.3.4 Social Attack

In a social attack, attackers send spam messages or mail to the users and try to steal their information. Spam filtering and fake social network detection techniques are used to detect the spoofed mail to avoid the attack.

9.5.4 Attacks on Data Integrity

9.5.4.1 Illusion Attack

Illusion attacks are when the attackers give the fake and useless information to the nodes about weather, traffic conditions, or accidents. This attack can divert a vehicle from their route, cause traffic jams, or cause accidents. To overcome such types of attacks, a plausibility validation network (PVN) model is used. This model gives a data check model which transmits data once it is verified to avoid an illusion attack. Figure 9.12 shows an illusion attack.

9.5.4.2 Masquerading Attack

In this attack, the vehicle is impersonated by another vehicle. The attacker creates false information about the vehicle to get access to personal information of the vehicles through acceptable access. The received data is analyzed in multiple layers to avoid the masquerading attack.

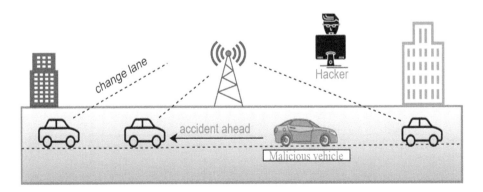

FIGURE 9.12 Illusion attack.

9.5.4.3 Replay Attack

In this attack, the same information is replayed again to give false information about the vehicle. The attackers stored the message received or sent by the other member vehicles. The malicious content is added in the stored message and it is again replayed by the attackers which gives the false information about the vehicle's position or velocity and eventually causes a hazardous problem.

9.5.5 Attacks on Non-Repudiation

9.5.5.1 Repudiation Attack

In repudiation attacks, a vehicle's node denies the transmission of data when it acts as a sender and reception of any data when it acts as receiver. So retransmission of data has to be done by the sender, which overuses network bandwidth and also consumes VANET resources. Figure 9.13 shows an attack on repudiation.

9.6 CHALLENGES AND FUTURE PERSPECTIVES

Some work should be done, taking into consideration the challenges and characteristics of VANETs given below, to have a new effective approach of communication.

9.6.1 Network Management

Because of the higher mobility, network topologies and channel conditions fluctuate rapidly. The usage of structures like trees suffers due to the quick changes in the topology, which makes them difficult to set up and maintain.

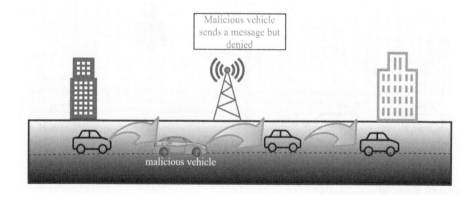

FIGURE 9.13 Attack on repudiation.

9.6.2 Congestion and Collision Control

There arises another challenge due to the unbounded network size. At night, both rural and urban areas have low traffic loads. In rural areas, traffic loads are low during the day and during the night in the urban areas. Thus, in the busy hours the traffic load is high and results in the occurrence of network partition frequently and results in network collision and congestion.

9.6.3 Environmental Impact

Due to their use of electromagnetic waves for communication, VANET deployment can cause environmental issues. Environmental factors can affect these waves.

9.6.4 MAC Design

A VANET typically uses a shared communication medium for communication, which means that MAC design is a major concern. The CSMA-based MAC used in IEEE 802.11 protocol for VANET is replaced by many techniques, including TDMA, SDMA, and CSMA, etc.

9.6.5 Security

VANETs provide many road safety applications that are in fact life critical; thus, the security of the messages must be ensured to preserve these messages. In VANETs, safety-related message needs to be passed on with a transmission delay of 100 ms, as these are time critical which actually makes it a real time constraint. Therefore, if real-time cryptography is to be achieved, fast algorithms should be introduced into use. There is a need of authentication of the message and entity as well in time.

9.6.6 Data Consistency Liability

Even authenticated nodes are at risk of being maliciously involved in VANET activities, causing accidents and disruptions within the network. Hence, there arises a need for a mechanism to avoid this inconsistency. This inconsistency can be avoided by looking for a correlation between data received from different nodes.

9.6.7 Low Tolerance for Error

The bases of design for some protocols can be probability. Using VANET, critical information is available in very short time frames to allow action to be taken quickly. In a probabilistic algorithm, even a small error could cause harm.

9.6.8 Key Distribution

Keys are essential to the functioning of all security systems in VANETs. Every message must be decrypted by the receiver with the same or a different key, and then it can be viewed. Even different ways can be used by different manufacturers to install keys. Thus, in designing security protocol, the key distribution among vehicles is a big concerning challenge.

9.6.9 Incentives

Apps that satisfy the consumer's needs and likes are always favored by manufacturers. Vehicles that can automatically report violations of traffic rules are unpopular among consumers. To achieve the successful deployment of such vehicular networks, vehicle manufacturers, consumers, and governments will need to be provided with incentives, which ultimately lead to concerns regarding the security implementation of VANETs.

9.6.10 Highly Heterogeneous Vehicular Networks

The rapid development of wireless networking has given rise to new non-interoperable technologies, and mobile computing has become an increasingly attractive option for consumers. As a consequence, we are currently experiencing a very complex interconnection among wireless technology platforms. If your concern involves node addressing, routing, security, quality of service, billing, or billing for quality of service, they may be involved. Therefore, in terms of network solutions, the next generation of intelligent transportation systems can be envisioned as a more holistic approach. As a result, broadband services would have to be made available through the coexistence of different wireless networks that are co-located close to each other.

9.6.11 Data Management and Storage

Millions of vehicles will constitute large-scale vehicular networks, resulting in a vast amount of distributed data. Such types of data must be stored and distributed properly across the VANETs. This feature relates to a huge network and a large amount of generated data, which in return will has its own unique challenges to cope with for data management.

9.6.12 Localization Systems

Highly accurate and reliable localized systems are required by VANETs for critical safety applications. By embedding GPS devices in all vehicles, a localization system for VANETs might be a natural fit. Nevertheless, satellite-based positioning systems may possess some undesirable traits like never being available 24 hours a day. Also, such systems are really bad to some kinds of attacks like blocking and spoofing. Furthermore, as a result, the VANET applications do not meet the needs for critical applications due to location errors of 10 feet to 30 feet, and other techniques would be more appropriate.

9.6.13 Disruptive Tolerant Communications

in a sparse network, issues like higher delay and low reliable delivery are of much concern. As a result of the carry-forward technique, you can increase the reliability of the delivery, but it will increase the time it takes for information to be delivered. Many novel approaches are available for data communication, such as using varied technologies for heterogeneous vehicular networks, to resolve or minimize the aforementioned issues. Additionally, driving behavior can be considered to enhance communications and reduce delivery times for information.

9.6.14 Tracking a Target

In a VANET, communication relies on the physical location of the vehicle, since that's the fundamental concern in any network. A VANET is fundamentally intended to track the target of its protocols, services, and applications that need such information. Using a tracking mechanism, nodes can be identified by the path they take in the network, as well as predicted by the network for their next location.

9.7 CONCLUSION

In summary, safety from attackers is the main attention for the VANET users. VANETs face several safety-related challenges of passengers' comfort and safety warnings for the drivers. The attackers interrupt the communication and try to get the information of communication systems. They are doing so to reduce the performance of a VANET communication system. The main reason behind more attacks in a VANET is because VANETs do wireless communication. So, it is easy for attackers to attack a VANET network. The detailed explanation of VANET attacks and their classifications like non-repudiation, authentication, availability, confidentiality, and data integrity are discussed in this chapter. Also, the possible challenges and security services are highlighted.

ABBREVIATIONS

RSU: Road Side Unit
GPS: Global Positioning System
DOS: Denial of Service

TCP:	Transmission Control Protocol
MAC:	Medium access control
PVN:	Plausibility Validation Network
TDMA:	Time-division multiple access
SDMA:	Space-division multiple access
CSMA:	Carrier-sense multiple access
VANETs:	Vehicular ad-hoc networks
VLC:	Visible light communication
V2V:	Vehicle-to-Vehicle
V2I:	Vehicle-to-Infrastructure

REFERENCES

[1] A. Quyoom, R. Ali and D. N. Gouttam, "A Novel Mechanism of Detection of Denial of Service Attack (DoS) in VANET using Malicious and Irrelevant Packet Detection Algorithm (MIPDA)," in Proceedings of the IEEE International Conference on Computing, Communication and Automation (ICCCA2015), pp. 414–419, 2015.

[2] T., Karimireddy and A., Bakshi, "A Hybrid Security Framework for the Vehicular Communications in VANET," in Proceedings of the International Conference on Wireless Communications, Signal Processing and Networking (WiSPNET), pp. 1929-1934, 2016

[3] K. M. A. Alheeti, A. Gruebler, and K. D. Mcdonald-Maier, "An Intrusion Detection System Against Malicious Attacks on the Communication Network of Driverless Cars," IEEE Consumer Communications and Networking Conference, no. 12, pp. 916–921, 2015.

[4] S. Zeadally, R. Hunt, Y.-S. Chen, A. Irwin, and A. Hassan, "Vehicular ad hoc networks (VANETS): status, results, and challenges," *Telecommunication Systems*, vol. 50, no. 4, pp. 217–241, 2012.

[5] V. Bibhu, "Performance Analysis of Black Hole Attack in Vanet," *International Journal of Computer Network and Information Security*, vol. 11, no. October, pp. 47–54, 2012.

[6] S. M. Safi, A. Movaghar, and M. Mohammadizadeh, "A novel approach for avoiding wormhole attacks in VANET," 2nd International Workshop on Computer Science and Engineering WCSE 2009, vol. 2, pp. 160–165, 2009.

[7] Wireless Communications, Signal Processing and Networking (WiSPNET); 2016: 1929-1934. 119. Ibrahim S, Hamdy M, Shaaban E. A proposed security service set for VANET SOA. Paper presented at: 2015 IEEE Seventh International Conference on Intelligent Computing and Information Systems (ICICIS); 2015:649-653.

[8] J. Hortelano, J. C. Ruiz, P. Manzoni. Evaluating the usefulness of watchdogs for intrusion detection in VANETs. Paper presented at: 2010 IEEE International Conference on Communications Workshops, pp. 1–5, 2010.

[9] S. Ruj, M. A. Cavenaghi, Z. Huang, A. Nayak, Stojmenovic I.. On data-centric misbehavior detection in VANETs. Paper presented at: 2011 IEEE Vehicular Technology Conference (VTC Fall), pp. 1–5, 2011.

[10] K. M. A. Alheeti, A. Gruebler, McDonald-Maier K. D.. On the detection of grey hole and rushing attacks in self-driving vehicular networks. Paper presented at: 2015 7th Computer Science and Electronic Engineering Conference (CEEC), pp. 231–236, 2015.

[11] M. N. Mejri, J. Ben-Othman Detecting greedy behavior by linear regression and watchdog in vehicular ad hoc networks. Paper presented at: 2014 IEEE Global Communications Conference, pp. 5032–5037, 2014.

[12] S. Suri Different methods and approaches for the detection and removal of wormhole attack in MANETS. *International Journal of Engineering Research & Technology*, vol. 1, no. 5, pp. 14–18, 2013.

[13] R. S. Bali, N. Kumar Secure clustering for efficient data dissemination in vehicular cyber–physical systems. *Future Generation Computer Systems*, vol. 56, (Supplement C), pp. 476–492, 2016.

[14] Z. Sun, Y. Liu, J. Wang, W. Deng, S. Xu Non-cooperative game of effective channel capacity and security strength in vehicular networks. *Physics Communications*, vol. 25, pp. 214–227, 2017.

[15] L. Karim, B. A. Bensaber, M. Mesfioui, I. Biskri A probabilistic model to corroborate three attacks in vehicular ad hoc networks. Paper presented at: 2015 IEEE Symposium on Computers and Communication (ISCC), pp. 70–75, 2015.

[16] H. Sedjelmaci, S. M. Senouci, M. A. Abu-Rgheff An efficient and lightweight intrusion detection mechanism for service-oriented vehicular networks. *IEEE Internet of Things Journal*, vol. 1, no. 6, pp. 570–577, 2014.

[17] O. S. Oubbati, A. Lakas, F. Zhou, M. Güneş, N. Lagraa, M. B. Yagoub Intelligent UAV-assisted routing protocol for urban VANETs. *Computer Communications*, vol. 107, (Supplement C), pp. 93–111, 2017.

[18] M. Conti, N. Dragoni and V. Lesyk, "A Survey of Man In The Middle Attacks," in *IEEE Communications Surveys & Tutorials*, vol. 18, no. 3, pp. 2027–2051, third quarter, 2016.

[19] S. Gupta, A. Singhal and A. Kapoor, "A literature survey on social engineering attacks: Phishing attack," 2016 International Conference on Computing, Communication and Automation (ICCCA), Greater Noida, India, pp. 537–540, 2016. doi: 10.1109/CCAA.2 016.7813778

[20] H. Saljooghinejad, Bhukya W. N., *Layered Security Architecture for Masquerade Attack Detection.ǁ Lecture Notes in Computer Science*, vol. 7371. Springer, Berlin, Heidelberg, 2012.

[21] N. Lo and H. Tsai, "Illusion Attack on VANET Applications - A Message Plausibility Problem," 2007 IEEE Globecom Workshops, Washington, DC, USA, pp. 1–8, 2007. doi: 10.1109/GLOCOMW.2007.4437823

[22] J. Li, H. Lu and M. Guizani, "ACPN: A Novel Authentication Framework with Conditional Privacy- Preservation and Non-Repudiation for VANETs," in *IEEE Transactions on Parallel and Distributed Systems*, vol. 26, no. 4, pp. 938–948, April 2015. doi: 10.1109/TPDS.2014.2308215

[23] H. H. R. Sherazi, R. Iqbal, F. Ahmad, Z. A. Khan, M. H. Chaudary Ddos attack detection: A key enabler for sustainable communication in internet of vehicles. *Sustainable Computing: Informatics and Systems*, 2019.

[24] R. Kumar, X. Zhang, W. Wang, R. U. Khan, J. Kumar, and A. Sharif, A multimodal malware detection technique for Android IoT devices using various features, *IEEE Access*, vol. 7, pp. 64411–64430, 2019.

[25] S. G., Hymlin Rose, & T., Jayasree, *Detection of jamming attack using timestamp for WSN*. Ad Hoc Networks, pp. 101874, 2019.

[26] B. Cherkaoui, A. Beni-Hssane, M. Erritali Variable control chart for detecting black hole attack in vehicular ad-hoc networks. *Journal of Ambient Intelligence and Humanized Computing*, vol. 11, pp. 1–10, 2020.

[27] L. Chen, K. A. Almoubayed, J. Leneutre Detection and prevention of greedy behavior in ad hoc networks. In: International conference on risks and security of internet and systems (CRISIS 2007), 2007.

[28] D. Zhang, Hui, Ge, T. Zhang, Y. Y. Cui, X. Liu, & G. Mao, New Multi-Hop Clustering Algorithm for Vehicular Adhoc Networks. *IEEE Transactions on Intelligent Transportation Systems*, vol. 20, pp. 1517–1530, 2018.

[29] A., Aldegheishem, H., Yasmeen, H., Maryam, M. A., Shah, A., Mehmood, N., Aleajeh, & H., Song, Smart road traffic accidents reduction strategy based on intelligent transportation systems. *Sensors*, vol. 18, no. 7, pp. 1983, 2018.

[30] J. M. Noman et al., "VANET's Security Concerns and Solutions: A Systematic Literature Review," in *Proceedings of the3-rd International Conference on Future Networks and Distributed Systems* (ICFNDS) ACM, pp. 1–12, July 1–2, 2019.

[31] J. Liu, Y. Yu, Y. Zhao et al., "An efficient privacy preserving batch authentication scheme with deterable function for VANETs," in *Proceedings of theInternational Conference on Network and System Security*, pp. 288–303, 2018.

Energy-Efficient Approximate Multipliers for ML-Based Disease Detection Systems

10

Zainab Aizaz and Kavita Khare
Department of Electronics and Communication Engineering, Maulana Azad National Institute of Technology, MANIT, Bhopal

Contents

10.1	Introduction to Machine Learning	204
10.2	Role of Machine Learning in Handling Biomedical Data	205
10.3	State-of-the-Art Machine Learning–Based Prediction Circuits	205
10.4	Efficient Approximate Multiplier for Neural Network–Based Disease Prediction Circuits	211
	10.4.1 The Booth Multiplier	211
	10.4.2 The Approximate Multiplier	213
	10.4.3 Characterization of Errors	214
	10.4.4 The Proposed Approximate Multiplier	214
10.5	Results	217
10.6	Applications on a Neural Network Case Study	218
10.7	Conclusion	221
References		222

10.1 INTRODUCTION TO MACHINE LEARNING

Machine learning is a field of artificial intelligence in which the systems learn automatically from their own experience (Goodfellow, Bengio & Courville, 2016). There are broadly three categories of machine learning algorithms: supervised, unsupervised, and reinforcement learning. Supervised learning use labeled information for future prediction while unsupervised learning uses unlabeled or unclassified information. Reinforcement learning interacts with its own environment and can produce errors or rewards, which allows them to recognize and adapt to the ideal behavior. The most popular ML algorithms are artificial neural networks (ANNs), and will be referred to simply as neural networks (NNs) throughout the text. NNs can be either supervised or unsupervised. NNs are networks of thousands of artificial neurons connected densely in the layers shown in Figure 10.1(a). The function of the artificial neuron is analogous to the biological neuron in the human brain. Each connection in Figure 10.1(a) has a weight associated with it. When data propagates in the forward direction only, it is termed a feed-forward neural network. It is used to calculate output vector from an input vector. Back propagation is essential in neural network training, as it tunes the

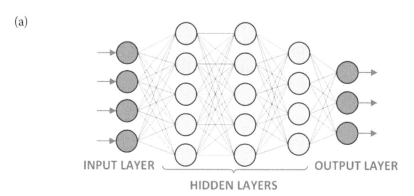

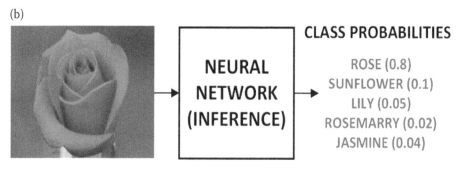

FIGURE 10.1 (a) A fully connected neural network; (b) NN used for classification of flower class.

weights after checking the error from the previous iteration. These finely tuned weights form a trained neural network model. The better the training, the more accuracy at the final output of the neural network.

10.2 ROLE OF MACHINE LEARNING IN HANDLING BIOMEDICAL DATA

Whether it is a wearable healthcare monitoring device or system-on-chip (SoC) for analysis, prediction, prognosis, or treatment of a disease, NNs stand at the nexus of artificial intelligence to provide promising and real-time solutions. This is due to the two outstanding abilities of NNs which are especially suited for disease applications, i.e., classification and clustering. Classification is useful for pattern recognition and therefore can easily classify abnormal cells or abnormal behavior of input signals that cause the disease or an unhealthy condition. The clustering property of NNs is used for feature selection and helps to identify variants of the disease. When the disease applications of NNs seem to be far-fetched (Mantzaris, Anastassopoulos & Lymberopoulos, 2008), a multilayer perceptron neural network (MLPNN) predictor for osteoporosis diagnoses was designed in the medical informatics laboratory, Greece. This device was aimed to save the patients from the harmful X-ray investigation or bone densitometry test for the diagnoses. The clinical parameters used for the NN training were age, sex, height, and weight. Since then, there has been a tremendous increase in research in NN-based biomedical circuits and there is a multitude of research papers on AI techniques focusing on the development of more efficient predictive solutions. Some examples are heart disease, Alzheimer's disease (Nguyen, He, An, Alexander, Feng & Yeo, 2020), renal disease (Vásquez-Morales et al., 2019), chest diseases (Salehinejad, Colak, Dowdell, Barfett & Valaee, 2019), all of which achieved a classification accuracy in the range of 70% to 95%. Inspite of the huge database available in the NN-based disease prediction, a paucity exists in the available directions to select the correct NN algorithm for a particular disease.

10.3 STATE-OF-THE-ART MACHINE LEARNING–BASED PREDICTION CIRCUITS

A number of disease prediction circuits based on prediction and diagnoses by machine learning algorithms have been designed. Electrocardiogram (ECG) and electroencephalogram (EEG) signals have a widespread utility for assessing the functionality of heart and brain. For implementing the biomedical circuits for ECG and EEG signal analysis, a number of dedicated hardware circuits are proposed in the last decade. For a complete biomedical circuit, it is essential that the hardware and software designs work synergistically. For the classification applications, selection of a proper algorithm is crucial and some of the algorithms used in the last decade for EEG and ECG signal monitoring are depicted in Table 10.1. Neural networks and support vector machine (SVM) (R. J. Martis et al., 2012) were used to classify the beats of an ECG signal. Finally, principal component analysis was used and accuracy of 85% was achieved. Similarly, SVM was also used with heuristic particle swarm optimization (PSO) algorithm and achieved accuracy of 91%, which implies that SVMs performance enhances when combined with heuristics. Simulation-based analysis of cancer was performed (L. Ohno-Machado and D. Bialek, 1998) using the back propagation algorithm of neural network back in 1998, while ECG monitoring using neural networks was performed in 1992 (Y. Ku, W. Tompkins, and Q. Xue, 1992). Both of these designs and many others propelled the requirement to design self-sufficient circuits for

TABLE 10.1 Summary of machine learning algorithms for biomedical applications

S. NO.	ALGORITHM	APPLICATION	REFERENCE
1.	Support Vector Machine (SVM)	ECG	• H. Khandoker, M. Palaniswami, and C. K. Karmakar, 2009 • G. Li and W.-Y. Chung, 2013 • Q. Li, C. Rajagopalan, and G. D. Clifford, 2014. • M. Bsoul, H. Minn and L. Tamil, 2011.
		EEG	• X.-W. Wang, D. Nie, and B.-L. Lu, 2011. • X. W. Wang, D. Nie, and B. L. Lu, 2014. • R. Panda, P. S. Khobragade, P. D. Jambhule, S. N. Jengthe, P. R. Pal, and T. K. Gandhi, 2010. • Y. Liu, W. Zhou, Q. Yuan, and S. Chen, 2012.
2	Convolutional neural network (CNN)	ECG	• U. R. Acharya et al., (a) 2017. • U. R. Acharya et al., (b) 2017. • U. R. Acharya et al., (d) 2017. • U. R. Acharya et al., 2018.
3.	Deep neural network (DNN)	EEG	• W. Zheng and B. Lu, 2015. • J. Suwicha, P. N. Setha, and I. Pasin, 2014.
4.	Random forest	EEG	• E. S. Pane, A. D. Wibawa, and M. H. Purnomo, 2019.
5.	Recurrent neural networks (RNN)	ECG	• U. E. Derya, 2009. • U. E. Derya, 2010.
6.	Feedforward and backpropagation neural networks	EEG	• T. Tzallas, M. G. Tsipouras, and D. I. Fotiadis, 2009. • S. M. S. Alam and M. I. H. Bhuiyan, 2013.

disease prediction, diagnoses, and treatment. At that time, graphic processing units (GPU) were used sparingly by only certain applications. As technology took a leap in the decade, more programmable circuits were invented such as application specific integrated circuits (ASIC) and field programmable gate array (FPGA) as stand-alone devices capable to implement ML applications. In the previous years, a multitude of deep learning–based biomedical circuits are proposed targeting numerous healthcare conditions, acute and chronic diseases, and circuits for general well-being (M. R. Azghadi et al., 2020). These circuits cover the entire range of functionalities including diagnoses, prediction, and treatment. Top companies such as Google and Nvidia have already built their graphic processing units (GPUs) and tensor producing units (TPUs) for fast training and testing of neural networks. Most recently, these companies are researching on application-specific integrated circuit (ASIC)–based biomedical circuits, because of their portability and low power consumption. Electromyographic signals (EMG) signals are useful for prosthesis control; a lot of research is focused on the hand gesture control classification (E. Ceolini et al., 2020). Wearable and portable low energy-delay-product (EDP) circuits are available for accurate monitoring and analysis of the prosthesis control mechanism. CMOS logic based Eyeriss is a convolutional neural network (CNN) accelerator (Y. Chen, T. Krishna, J. S. Emer and V. Sze, 2017), having low energy consumption, reconfigurable to adapt to various CNN architectures. It has high throughput due to its parallel data flow and row-stationary spatial arrangement of 168 elements for computation. Thyroidal cancer prediction is performed using Eyeriss (Q. Guan et al., 2019) and the CNN deployed for this application is VGG-16. Similarly, an investigation on different parameters of CNN training and fine tuning is presented for various medical imaging applications such as radiology, cardiology, etc. (N. Tajbakhsh et al., 2016). The research is performed on the Alexnet CNN deployed on Eyeriss.

For almost three decades, the preponderance of computer architecture designs exists on CMOS technology based integrated circuits (ICs). But there are three walls (N. Ibrahim, S. Salah, M. Safar and M. W. El-Kharashi, 2018) that are hindering further miniaturization and performance of CMOS ICs; these are first the memory wall that restricts the number of memory fetch and store, the parallelism wall that restricts completely parallel execution, and the power wall which restricts the further increment in the speed of the clock. Therefore, developing new logic devices using CMOS is approaching saturation, and therefore new devices and architectures are prerequisites of the VLSI industry. Memristive devices have some outstanding features which if pulled together on ICs can provide multifold gains in the hardware design. These features are firstly, the ability to provide negligible power on standby, secondly, very high density compared to other logics, thirdly, the compatibility with existing CMOS architectures, and finally the most amazing ability to structure itself in the form of memories that can facilitate in-memory computations. Memristive logic is recently taking over the CMOS logic and biomedical circuits based on the advantage that memristors can behave directly as a weight in DNN and spiking neural network (SNN) logic-based devices. Memristive circuits have another advantage in ANN applications that they have inherent ability to store logic, and they can be used for multiply-accumulate circuit inside the memory. This saves a large amount of energy required to store and fetch the values of neurons during millions of computations inside the ANN circuit. Memristive logic implemented using crossbars can achieve an energy saving of 2,500 times and speed increased by 25 times compared to the GPU and CPU alternatives. Memtorch (C. Lammie, W. Xiang, B. Linares-Barranco, and M. R. Azghadi, 2020) can be called a memristor-based substitute of the Eyeriss accelerator. The designers of Memtorch have implemented crossbars to facilitate the in-memory computation. Highly efficient multiplications and unrolling operations can be performed. One drawback of memristive devices is the early aging and non-ideal behavior. For implementing the neural network using memristors, pretrained neural networks are formed using crossbars. A single crossbar contain one transistor and one resistor and the memristor can be modeled by the properties of any of the standard memristive models (Yang, J., Strukov, D., & Stewart, D., "Memristive devices for computing". Nature Nanotechnology, 8, 13–24, 2013). Recently, cardiac arrythmia (A. M. Hassan, A. F. Khalaf, K. S. Sayed, H.H. Li, and Y.Chen, 2018) investigation is performed using memristive neural network. More than 6,000 real-time samples are used as a data set. The training is done using MATLAB, and memristors are programmed based on conductance values. The conductance values for on and off memristor are stored in the program and are retrieved at the time of simulation. For this design, a hardware circuit is not created and the whole design is simulation based. A Memristive network has not yet developed enough to implement hardware-based complex deep learning networks required for temporal and complex images in medical applications. Before verifying the effectiveness of the memristive neural networks, complete hardware implementation is required and therefore, Memristive neural networks (F. Cai et al., 2019) have not been used stand-alone in the biomedical applications. In order to achieve low power operation, for example, in case battery operated biomedical circuits, spiking neural networks (SNNs) can be used. SNN-based neuromorphic computing has two very important features, namely very low power consumption and real-time data processing. Only a few neuromorphic computing-based biomedical processors are available in literature, but they demonstrate energy and speed gains that outperform the hardware performances of the CMOS counterparts.

Figures 10.2, 10.3, and 10.4 show a few examples of biomedical circuits designed using different ML algorithms and using specific technologies such as CMOS, Memristors, and FPGAs. CMOS circuits represent the ASIC design in which a stand-alone user defined IC is designed and which is dedicated entirely for the execution of the particular application. Due to dedicated nature, their development cycle is long and very costly, from design on paper to the circuit in hand there is a huge barrier between costs of ASICs and FPGAs. At this cost, there are several prominent benefits that have made ASICS still the most favorable choice for IC consumers. The benefits include very high power and energy efficiency, low delays, and most importantly the ability to conglomerate analog and digital designs on the same die. FPGAs are programmable circuits, and are reconfigurable. They have high-throughputs and are used efficiently for prototyping of ASICs. One major drawback of FPGAs is the inability to include an analog circuit or a block on it. It is entirely dedicated to digital design (A. M. Sodagar, 2014). Due to this feature, the biomedical circuit design on an FPGA is not compact and rather spread over a large area as the signal processing units and transducers have to be reconfigured to work alongside FPGA and cannot be embedded onto FPGA. Digital circuits alone cannot be combined to create a complete biomedical circuit;

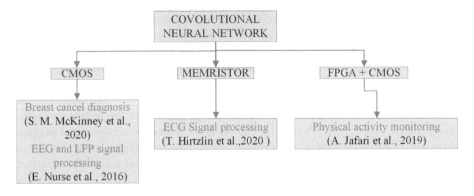

FIGURE 10.2 Biomedical circuits based on convolutional neural networks.

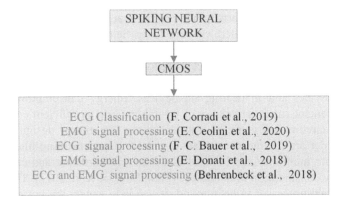

FIGURE 10.3 Biomedical circuits based on spiking neural networks.

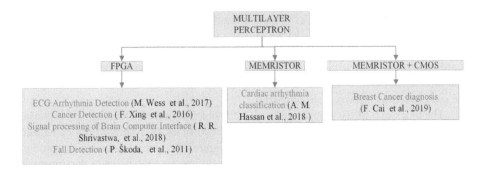

FIGURE 10.4 Biomedical circuits based on multilayer perceptron neural networks.

rather, a biomedical circuit includes signal converting and recording circuits, signal amplifiers, display units, and storage devices. Therefore, biomedical circuit designers use FPGA for simulation and classification tasks and majorly for prototyping the final design on ASIC counterpart.

Freezing of gait is an abnormal condition for the patient of Parkinson's disease in which the patient cannot move or walk even though he intends for it. This may let the patient fall because the torso is moving but the legs are stiff. A real-time, FPGA-based, wearable, freezing of gait classifier is designed (Mikos et al., 2018) that can be worn on the patients' limbs. An inertial measurement circuit transfers the signals obtained from limb movements to the processor. The processor on Artix-7 Field programmable gate array (FPGA) (XC7A35T) is constructed in two parts, one is the feature extraction circuit and the other is the NN core for classification. Three features are used as inputs to the feature extraction circuit are variance, stride peak, and freeze index. The freeze index is calculated using discrete Fourier transform of the input waves obtained from the body.

The calculation of 128-point DFT requires 384 (3N) adders and 256 (2N) multipliers, to be realized using FPGA DSP blocks. For the calculation of variance and stride peak, the angular velocity signals from the inertial measurement circuit are used. The NN classifier starts by writing the features into RAM parallelly, then the induced fields are calculated, passing these fields through sigmoid activation function programmed in the look up table (LUT) and finally the classified output is obtained. Among the wearable devices, speech signal processors have gained significant attention due to their application in blind/deaf language learning-based biomedical devices and in hearing-aid devices. Energy consumption has been a crucial requirement since the genesis of wearables. Energy efficient components on the wearable circuits is a common way to achieve low energy designs. Since multiplier and adder are the most important circuits for implementing filter operations, an approximate multiplier can be used that uses less hardware and can reduce the overall energy consumption of the device (Park, Shin, & Kim, 2017). The major goal of speech signal processor in hearing-aid device is to cancel the noise. A feedback noise-canceling algorithm can be implemented using a finite impulse response (FIR) filter (Bonetti, Teman, Flatresse & Burg, 2017). An approximate multiplier can be used inside a FIR filter, as shown in Figure 10.5. If kth noisy input is $A(k)$, $\omega(k)$ is the filter weight vector and $v(k)$ is the feedback vector, the error is calculated using difference between $A(k)$ and $v(k)$. $O(k)$ is the output of the dynamic range selection circuit that modifies the output that is most accurate for the wearer. The approximate multiplier used in this circuit is based on truncated operands. Before the multiplication takes place, 5 bits out of 8 bits in the operand depending on the leading one bit are chosen as shown in Figure 10.6.

The block of 5 MSBs (most significant bits) which contain the leading 1 bit is chosen using a truncation window in both the operands and block of 10 bits MSBs containing the leading 1 bit chosen from the product. In Figure 10.5, let us calculate the product using both exact and approximate multiplication. Before the application of the truncation window, the exact operands are 95 (01011111_2) and 21(00010101_2) and the exact product is 1995(0000111100111000_2), but after applying the truncation window, the approximate operands are 88 (01011_2) and 21(10101_2) and the approximate product is 1848(0011100111_2). The error in the output is 7.3%. Due to a smaller number of bits in both operands and the product, the overall hardware circuitry of the multiplier is simplified and it becomes more energy efficient. In particular, for devices in wearable or

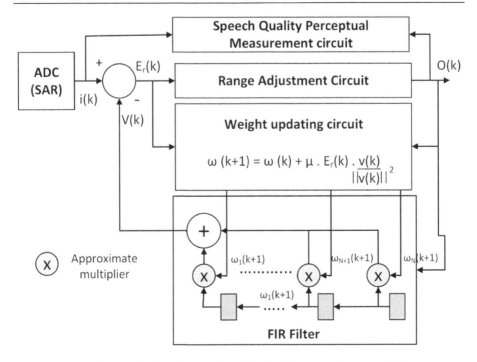

FIGURE 10.5 The speech signal processing circuit using approximate multipliers.

Multiplicand		0	1	0	1	1	1	1	1	$88_{(10)}$
Multiplier		0	0	0	1	0	1	0	1	$21_{(10)}$
Product	0	0	0	0	1	1	1	1	0	0	1	1	1	0	0	0	$1848_{(10)}$

FIGURE 10.6 The bit selection technique of approximate multiplier.

portable handheld applications, the low energy consumption feature of this device is a significant improvement towards designing better circuits.

10.4 EFFICIENT APPROXIMATE MULTIPLIER FOR NEURAL NETWORK–BASED DISEASE PREDICTION CIRCUITS

10.4.1 The Booth Multiplier

In 1951, A. D. Booth first proposed a signed binary multiplication over a digital logic device by using the method of partial products (Booth, 1951). The general scheme for binary multiplication is similar to decimal multiplication. On a digital IC, the multiplier is designed using three steps. In the first step, the partial products are generated in which each partial product is generated by multiplying the multiplicand with a bit of the multiplier using AND operation (in case of AND-array multipliers). For N bit operands, a partial product matrix of n rows is generated. In the case of the Booth multiplier, the partial product matrix is generated by the Booth encoder circuit (Elguibaly, 2000). The advantage of applying the Booth algorithm is that the number of partial products is reduced to at least half (N/2), and therefore, it is extensively used in multipliers with long operands (>16 bits). The disadvantage of the Booth multiplier is that it is comparatively more complex due to the requirement of the recoding circuit. In the second step, the partial products are added column-wise using the Wallace tree or Dadda tree structure (Dadda, 1989). Booth multipliers are very commonly used due to their low power consumption and low critical path. Bit adding circuits are used in this accumulation process using half adder, full adder, and compressor circuits. In the final step, a final adder is required to generate the multiplication result as the number of partial products is reduced to sum and carry only. Carry look ahead adders (CLAs) are commonly used in this stage as they are faster compared to other adders. On a circuit, a multiplier is combined with adder to perform an important operation of multiply and accumulate. A MAC (multiply-accumulate) unit is composed of adder, multiplier, and an accumulator. In a processor, the inputs for the MAC are fetched from a memory location and fed to a multiplier block of MAC, which will perform multiplication and give the result to the adder that will accumulate the result and then will store the result in a memory location. FPGAs have digital signal processing (DSP) slices (DSP blocks) to implement signal processing functions (Gao, Khalili & Chabini, 2009). The DSP operation most commonly used is a MAC operation (Table 10.2).

In a Booth encoding circuit, three consecutive bits of the multiplier and two consecutive bits of multiplicand are used to derive the partial product rows using the logic shown in Table 10.1. Figure 10.7 demonstrates the process of Booth multiplication. A radix-m Booth algorithm uses n bits of the multiplier and two bits of multiplicand to encode the partial product bit where the value of n is $\log 2m + 1$. Therefore, the radix-2 algorithm uses 2 bits of multiplicand, radix-4 algorithm uses 3 bits of multiplicand, and the radix-8 algorithm uses 4 bits of multiplicand and so on.

TABLE 10.2 Radix-4 Booth algorithm

B_{I+1}	B_I	B_{I-1}	BOOTH ENCODING (PARTIAL PRODUCTS)
0	0	0	0 * Multiplicand
0	0	1	1 * Multiplicand
0	1	0	1 * Multiplicand
0	1	1	2 * Multiplicand
1	0	0	− 2 * Multiplicand
1	0	1	−1 * Multiplicand
1	1	0	−1 * Multiplicand
1	1	1	0 * Multiplicand

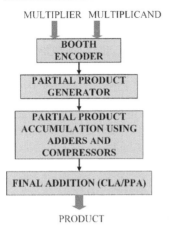

FIGURE 10.7 Block diagram of Booth multiplier.

10.4.2 The Approximate Multiplier

An approximate multiplier does not provide exact expected output of the multiplication operation. This inexactness is introduced into the multiplier by pruning its complete circuit or by employing pruned sub-circuits in the multiplier. These pruned sub-circuits are approximate Booth encoders to be used at partial product generation level, approximate full and half adders and compressor circuits to be used at the partial product addition level, approximate carry look ahead adders (CLAs) or carry prefix adders (CPAs) to be used at the final addition level. There are several other ways to insert approximation in a multiplier such as by using truncated, rounded, or encoded operands at the inputs (Vahdat, Kamal, Afzali-Kusha & Pedram, 2019). A very important approximation technique commonly employed is the truncation of least significant part of the multiplier partial product matrix, thereby pruning the circuit at the partial product generation level, partial product accumulation level, and at the final addition level. Since the partial products at the least significant portion are cut off, the error introduced is smaller compared to cutting off at the other parts of the partial product matrix. Suppose that six least significant columns of the partial product matrix are truncated, then a maximum difference of 63 ($2^5 + 2^4 + 2^3 + 2^2 + 2^1 + 2^0$) exists between exact and inexact output. By employing truncation, the whole circuit of the truncated part is not generated at all, thereby saving that particular amount of area. Due to truncation, power and delay are also reduced significantly. Despite of the hardware savings, a truncation error is induced in the multiplier. To reduce the error generated due to truncation, a number of error compensation techniques are proposed (Ha & Lee, 2018). An error compensation circuit is a small circuit used to mitigate or cancel the error generated in an approximate multiplier. Most of the state-of-the art compensation circuits are designed by using the partial products from the truncated part. Since the partial products are related to each other, a judiciously designed compensation circuit from the partial product combination can alleviate the error induced by truncation (Z. Aizaz and K. Khare, 2021). The error compensation circuits are mainly applied at the beginning of the non-truncated part or to the final addition part to produces significant error cancellation in the remaining part of the output of the multiplier.

10.4.3 Characterization of Errors

There are several well-established error metrics that are used to characterize the error generated due to applied approximation (Mahdiani, Ahmadi, Fakhraie & Lucas, 2010). The amount of error needs to be minimized for a satisfactory performance of any application. Therefore, comparing the error metrics of different designs can help to justify the trade-off between accuracy and power efficiency for approximate circuits. Essentially, these metrics determine the average error over the entire sample space of the circuit. The most appropriate and commonly used error matrices are listed in Table 10.3.

10.4.4 The Proposed Approximate Multiplier

Compressor circuits are used for a fast accumulation of the partial products in the multipliers. The other name for compressors is the "one counter." Since the emergence of approximate computing, there is a bulk of designs that focus on pruning the compressor circuits by manipulating the truth table to reduce the complexity of the circuit (Strollo, Napoli, De Caro, Petra & Meo, 2020). An exact 5-3 consists of 5 inputs and 3 outputs and uses two full adders with a gate count of 10 gates. In this brief, an approximate compressor circuit is designed consisting of 4 inputs and 2 outputs. The proposed compressor contains only six gates, thereby reducing the circuit area, delay, and power consumption. The technique used to design the approximate 4-2 compressor is K-map modification, as shown in Figure 10.8. In Figure 10.8(a), the K-map of exact compressor is shown while in Figure 10.8(b), the modified K-map and its mapping is shown that is used to design the circuit of approximate compressor, as shown in Figure 10.9. The bits shown in green are the ones that are flipped from 1 to 0 and vice versa. The equations for the sum (S) and carry (C) of the approximate compressor derived from the K-map are:

$$S = (c \cdot \odot \cdot d)(\bar{a}\bar{b}) \tag{10.1}$$

TABLE 10.3 Error metrics for approximate multipliers

ERROR METRIC	EXPRESSION
MRED (Mean relative error distance)	MRED = $2^{2N^1} \sum_{i=1}^{2^{2N}} |ED_i|/M_i$
NMED (Normalized mean error distance)	NMED = $2^{2N}(2^N - 1)^{2^1} \sum_{i=1}^{2^{2N}} |ED_i|$
NoEB (Number of effective bits)	NoEB = 2N-log2(1+RMSED)
ER (Error Rate)	ER = $\frac{\text{Total number of erraneous Outputs}}{\text{Total number of Outputs in an experiment}}$
PRED (Probability of RED less than a specified percentage, assumed to be 2% in (Liu, Qian, Wang, Jiang, Han & Lombardi, 2017))	ER = $\frac{\text{Total number of outputs with RED<2\%}}{\text{Total number of Outputs in an experiment}}$

ED = |M − A|, RED = ED/M, N = number of bits in the operand (ED = Error distance, M = Exact multiplier output, A = Approximate multiplier output, N = Number of bits in the operand, RMSED = Root mean square of error distance over entire input space).

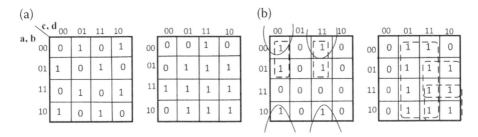

FIGURE 10.8 (a) K-map of the exact compressor; (b) modified K-map of the proposed approximate compressor.

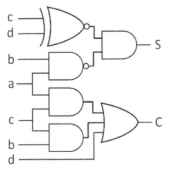

FIGURE 10.9 Circuit diagram of the approximate compressor.

$$C = d + c(a + b) \tag{10.2}$$

The proposed approximate compressor is applied at the least significant part of the Booth multiplier circuit, as shown in Figure 10.10. The proposed approximate compressor is applied from column 4 to column 16 and then the designs are classified based on some key characteristics such as the error metrics and the delay/power saving for the ASIC-based design. It is established that for the multiplication operation in a neural network, a precision of 8 bits is sufficient (Ansari, Mrazek, Cockburn, Sekanina, Vasicek & Han, 2020), but keeping in view the meticulous requirements of disease applications, a precision of 16 bits or more is superior for machine learning applications. Therefore, we have used 16-bit signed multiplier for the NN applications.

APPROACH A: *Truncation at LS columns with exact compressors* – Truncating or removing the least significant columns is established as a viable technique for implementing approximate multipliers. An approximation factor or truncation factor, t is used to create a range of truncated designs, suited for customized applications. The multiplier is truncated from length 1 to 18. By analyzing the designs on MATLAB, we confirmed that beyond the 18th column, the accuracy decreased drastically and the results are no more usable by any application. Exact compressors have extra carry output as shown in Figure 10.10, this carry increases the critical path of the circuit and therefore, the delay increases.

216 Advanced Circuits and Systems for Healthcare and Security Applications

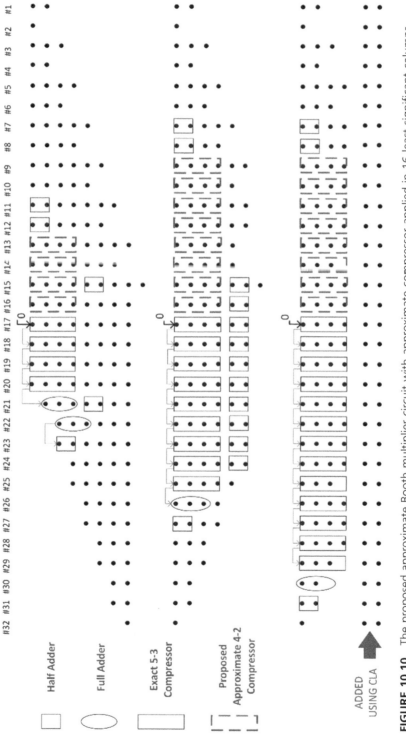

FIGURE 10.10 The proposed approximate Booth multiplier circuit with approximate compressor applied in 16 least significant columns.

APPROACH B: *Truncation with proposed approximate compressors–* The proposed approximate compressor is applied to the 16 least significant columns and the effects of truncation are observed as shown in Table 10.5. For the truncation length of 4 columns from the right, the proposed approximate compressor is applied from 5th column to the 16th column. In general, when t columns are truncated the approximate compressor is used from (t+1)th column up to the 16th column. Since the approximate compressor itself introduces error in the multiplier circuit, they are not used beyond half the length of the product (N/2). Experiments conducted beyond 16th column produced large errors in the output of the multiplier. Therefore, in order to preserve the accuracy of the output, experimentation is performed up to column 16.

10.5 RESULTS

The proposed approximate compressor-based Booth multipliers are first described in Verilog HDL at gate-level and verified by using Synopsys VCS. Verification of design is performed using 100,000 inputs to the test bench and then observing the waveform. The designs are then synthesized using synopsys design compiler at Synopsys 90 nm SAED technology library using the typical corner. The toggle frequency used is 1 GHz with the operating voltage of 1.2 V. The temperature used is 25°C. The power delay product is an important metric in the digital signal design as it indicates the amount of energy used by the design when it is operated with a set of applied input voltage. The PDP values of our design are comparatively less, which signifies that our design improves the energy consumption of the device. In literature, several biomedical circuits show the average in the range of several hundreds of mW (M. R. Azghadi et al.). Our approximate multiplier is an energy-efficient design that reduces the overall power consumption of the processor it is used with. Table 10.4 shows an improvement

TABLE 10.4 Comparison of proposed approximate multiplier with the existing multipliers.

DESIGN	AREA (MM^2)	POWER (MW)	DELAY (NS)	PDP (MW.NS)	MRED (%)	NMED (10^{-5})	PRED (%)	NOEB
Exact Booth	10190	5437	5.27	28653	–	–	–	–
Proposed	8990	4690	4.36	20448	0.35	3.61	98.76	16.42
Momeni, Han, Montuschi, Lombardi, 2015	9631	4981	4.36	21717	0.68	1.31	98.72	15.89
Ha & Lee, 2018	9616	4713	4.57	21538	0.54	0.65	98.89	16.13

TABLE 10.5 The variation in mean square error of a neural network with successively truncated approximate Booth multiplier

DESIGN				MSE			
EXACT				0.016290			
EXACT (16BIT)				0.016992			
APPROACH A				APPROACH B			
DESIGN T	MSE	DESIGN T	MSE	DESIGN T	MSE	DESIGN T	MSE
1	0.016990	10	0.017337	1	0.116138	10	0.119120
2	0.016993	11	0.017734	2	0.116636	11	0.119591
3	0.016998	12	0.019187	3	0.117156	12	0.120084
4	0.017007	13	0.023841	4	0.117899	13	0.124613
5	0.017014	14	0.033447	5	0.118008	14	0.126892
6	0.017032	15	0.083456	6	0.118027	15	0.132562
7	0.017050	16	0.110581	7	0.118037	16	0.135621
8	0.017042	17	0.166922	8	0.118059	17	0.200456
9	0.017096	18	0.169701	9	0.118321	18	0.208793

of 12%, 17.3%, and 13.7% in terms of area, delay, and power, respectively, by the proposed approximate multiplier over the exact Booth multiplier. For the analysis of error metrics, the proposed multipliers are designed in MATLAB, and the metrics are calculated using 10 million input vectors. The results for the proposed multipliers are summarized in Table 10.3. The accuracy of proposed design is high and it can be observed that it provides better MRED, PRED, and NoEB values compared to other state-of-the-art designs (Momeni, Han, Montuschi, Lombardi, 2015; Ha & Lee, 2018). The proposed 4-2 compressor is designed in such a way that the hardware and the accuracy of the proposed compressor are altered in a balanced way giving rise to an energy efficient design with improved accuracy. Figure 10.11 shows the simulation waveform of the exact and approximate multiplier outputs, Figure 10.11(a) shows 1 × 100 = 100 but Figure 10.11(b) clearly indicating inaccurate results i.e. 1 × 100 = 0. As described already, that approximate circuit has inaccurate outputs but at the same time provides large gains in area, power, and delay performances.

10.6 APPLICATIONS ON A NEURAL NETWORK CASE STUDY

Urbanization and women's fertility are correlated. Factors such as smoking, age, diseases, body weight, eating disorders, radiation, chemotherapy, and genetic factors contribute to the infertility in both men and women. Recently, due to increase in

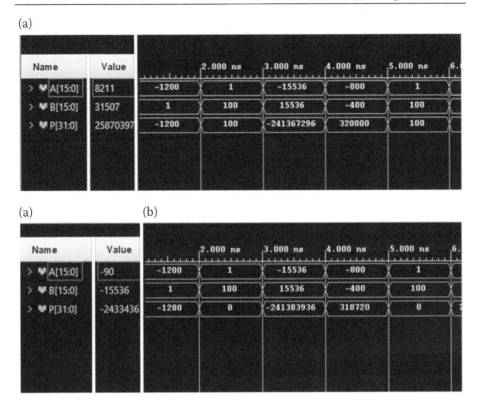

FIGURE 10.11 (a) Simulation waveform of exact multiplier; (b) results of approximate multiplier (first and second waveforms are multiplier and multiplic while the third waveform is for the product).

urbanization, female infertility increases has taken up arms, up to 12% due to radiation and smoke filled environment and up to 11% due to overweight or underweight (A.M.H. Koning, W.K.H. Kuchenbecker, H. Groen, A. Hoek, J.A. Land, K.S. Khan, B.W.J. Mol, 2010). Due to chemotherapy, premature failure of ovaries occurs and infertility is induced in women (S. Morgan, R.A. Anderson, C. Gourley, W.H. Wallace, N. Spears, 2012). The problem that we are analyzing is for the classification and prediction of fertility for becoming pregnant in urban women. The parameters used in the data set are season, age, diseases at time of birth, accidental trauma or depression, previous caesarean, alcohol consumption, smoking, and activity per day. The value of the output neuron can be 0 or 1 depending on whether the input case is fertile or not. A fully connected neural network with 9 neurons in the input layer, 7 neurons in the hidden layer, and 1 neuron in the output layer is designed in MATLAB. The original data set is split into two sub-sets, one for training containing 90% of the data and the other for testing containing 10% of the data. For assessing the accuracy of the neural network, the splitting of a data set is an important step. The approximate multiplier is used to multiply input with randomly generated synaptic weights. During the training process, the data is first fed to the network, then back propagation takes place, the errors are measured and synapses are measured to minimize the error, called the gradient descent. The gradient descent is the most commonly used algorithm in the neural network training. Its goal is to adjust the weights for minimizing any deviation from the correct output. The learning rate α is 0.001, which is important for proper convergence of the back propagation algorithm. Figure 10.12 shows the variation in error at the different values of iteration during the training phase. The total number of iterations used are 600,000. The mean square error (MSE) is an important metric in the context of neural networks; it is defined as the difference between desired output and the output of the network at a particular iteration. The mean square error for the trained data is reported in Table 10.4 for both the approaches A and B. In Figure 10.12, the reduction in error with iterations is demonstrated, showing the behavior of the back propagation algorithm during the training. In Table 10.4, the final MSE for different values of truncation factor are shown. It is not surprising to observe that as t increases, the MSE increases due to gradual rise in inaccuracy of the multiplier after every truncation. Table 10.5 shows the MSE obtained by using exact multiplication, which is the normal multiplication operation used by the CPU at IEEE-754, 64-bit floating point standard.

But exact (16-bit) uses low precision as the 32-bit multiplier output is divided into fixed-point 16-bit integer part and 16-bit fractional part. An increase of 4.3% is observed in MSE using 16-bit fixed point Booth multiplier as compared to 64-bit floating point standard. In approach A, we simply truncated the 16-bit Booth multiplier successively and derived the values of the MSE, as shown in Table 10.5. Let us compare t = 5 and t = 13, an increase of 40.12% in MSE takes place. The value of MSE further increases when we used the proposed approximate compressor-based Booth multiplier. Since a compressor in Approach A is exact and a compressor in Approach B is approximate, the MSE of approach B is inherently higher. Comparing t = 5 and t = 13 for approach B, an increase of 5.6% in MSE takes place. This increase in MSE is not very large but the hardware gain achieved due to truncation of LSB columns from t = 5 to t = 13 is much higher (Osorio & Rodríguez, 2019). Therefore, the proposed design of approximate multiplier provides higher energy efficiency along with minimal loss of accuracy (Figure 10.13).

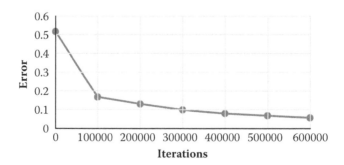

FIGURE 10.12 MSE during training of neural network.

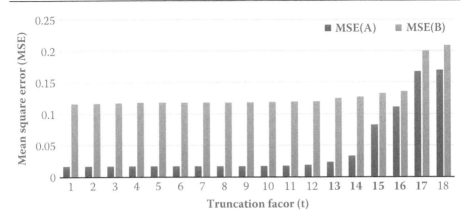

FIGURE 10.13 MSE of trained data for approach A (using truncated approximate Booth multiplier) and Approach B (using proposed approximate compressor in truncated Booth multiplier).

10.7 CONCLUSION

The accelerating research on large amounts of healthcare devices will intrinsically change the systems of medical care. It is our firm belief that the machine learning and healthcare relationship will become the keystone of the future medical management. The delivery of instant care to the patients will be assisted by low-power, portable, and high-speed efficient devices which will be the outcome of conglomeration of advanced VLSI design, machine learning, and advanced computing fields. Artificial intelligence has proven to be efficient for modeling, classification, and diagnoses of a multitude of diseases and healthcare problems. Previously, many computational algorithms had been used in the biomedical circuits like regression-based algorithms and evolutionary and heuristic algorithms but the drawback that limited them to the research domain and that they were not implemented practically in daily life was the lack of accuracy. An exceptional aspect of machine learning algorithms is that the output is very accurate due to which the biomedical circuits have crossed-over higher leaps in the field of prognosis, diagnoses, and treatment. An important drawback of digital biomedical circuits over conventional analog-digital ones is that when charges in battery decreases with prolonged use, the output loses its accuracy. For instance, a simple digital blood pressure monitor and a diabetes meter give ±10% less accurate output. This problem can be alleviated by using machine learning–based algorithms in the future. Therefore, there are innumerable ways machine learning can benefit the biomedical engineering area.

In this chapter, the relevance of machine learning for biomedical circuits is explained, and detail on different biomedical circuits using different IC technologies is provided. The chapter also provides an insight to the logic design of a novel compressor-based approximate multiplier. The approximate multiplier when used on circuits implementing neural networks for biomedical applications can provide large energy gains and increased speed. Efficient NN-based SoCs, ASIC- and FPGA-based NN accelerators, and wearable NN-based devices, all require low-power multiplication units as a basic unit of computation through the NN layers. An important requirement is low power, as some applications need to stand long hours of operation and at the same time for circuits that can be driven by battery for the wearable applications. Evidently, the proposed approximate multipliers can provide high energy efficiency with lower degradation in the quality of result. A range of truncation-based designs are proposed that can be used for several customized applications.

REFERENCES

I., Goodfellow, Y., Bengio, A., Courville, *Deep Learning*, MIT Press, 2016.

D. H. Mantzaris, G. C. Anastassopoulos and D. K. Lymberopoulos, "Medical disease prediction using Artificial Neural Networks," 2008 8th IEEE International Conference on BioInformatics and BioEngineering, Athens, Greece, pp. 1–6, 2008.

M. Nguyen, T. He, L. An, D. C. Alexander, J. Feng, B. T. Thomas Yco, "Predicting Alzheimer's disease progression using deep recurrent neural networks", *NeuroImage*, vol. 222, pp. 117203, 2020.

G. R. Vásquez-Morales, S. M. Martínez-Monterrubio, P. Moreno-Ger and J. A. Recio-García, "Explainable Prediction of Chronic Renal Disease in the Colombian Population Using Neural Networks and Case-Based Reasoning," in *IEEE Access*, vol. 7, pp. 152900–152910, 2019.

H. Salehinejad, E. Colak, T. Dowdell, J. Barfett and S. Valaee, "Synthesizing Chest X-Ray Pathology for Training Deep Convolutional Neural Networks," in *IEEE Transactions on Medical Imaging*, vol. 38, no. 5, pp. 1197–1206, May 2019.

I. Mikos et al., "A Neural Network Accelerator With Integrated Feature Extraction Processor for a Freezing of Gait Detection System," 2018 IEEE Asian Solid-State Circuits Conference (A-SSCC), pp. 59–62, 2018.

T. Park, K. Shin, and N. Kim, "Energy-Efficient Approximate Speech Signal Processing for Wearable Devices", *ETRI Journal*, vol. 39, no. 2, 2017.

A. Bonetti, A. Teman, P. Flatresse and A. Burg, "Multipliers-Driven Perturbation of Coefficients for Low-Power Operation in Reconfigurable FIR Filters," in *IEEE Transactions on Circuits and Systems I: Regular Papers*, vol. 64, no. 9, pp. 2388–2400, Sept. 2017.

Booth, A. D., 'Signed binary multiplication technique', *Quarterly Journal of Mechanics and Applied Mathematics*, 4, pp. 236–240, 1951.

F. Elguibaly, "A fast parallel multiplier-accumulator using the modified Booth algorithm," in *IEEE Transactions on Circuits and Systems II: Analog and Digital Signal Processing*, vol. 47, no. 9, pp. 902–908, Sept. 2000.

L. Dadda, "On serial-input multipliers for two's complement numbers," in *IEEE Transactions on Computers*, vol. 38, no. 9, pp. 1341–1345, Sept. 1989.

Gao, S., Al-Khalili, D., Chabini, N. "Efficient scheme for implementing large size signed multipliers using multi granular embedded DSP blocks in FPGAs", *International journal of Reconfigurable Computing*, vol. 2009, January, (1). Article ID 145130, 11 pages, 2009.

S. Vahdat, M. Kamal, A. Afzali-Kusha and M. Pedram, "TOSAM: An Energy-Efficient Truncation- and Rounding-Based Scalable Approximate Multiplier," in *IEEE Transactions on Very Large Scale Integration (VLSI) Systems*, vol. 27, no. 5, pp. 1161–1173, May 2019.

M. Ha and S. Lee, "Multipliers With Approximate 4–2 Compressors and Error Recovery Modules," in *IEEE Embedded Systems Letters*, vol. 10, no. 1, pp. 6–9, March 2018.

H. R. Mahdiani, A. Ahmadi, S. M. Fakhraie, and C. Lucas, "Bio-inspired imprecise computational blocks for efficient vlsi implementation of softcomputing applications," *IEEE Transactions on Circuits and Systems I: Regular Papers*, vol. 57, no. 4, pp. 850–862, Apr. 2010.

A. G. M. Strollo, E. Napoli, D. De Caro, N. Petra and G. D. Meo, "Comparison and Extension of Approximate 4–2 Compressors for Low-Power Approximate Multipliers," in *IEEE Transactions on Circuits and Systems I: Regular Papers*, vol. 67, no. 9, pp. 3021–3034, Sept. 2020.

Liu, W., Qian, L., Wang, C., Jiang, H., Han, J., Lombardi, F.: Design of Approximate Radix-4 Booth Multipliers for Error-Tolerant Computing, *IEEE Transactions on Computers*, vol. 66, no. 8, pp. 1435–1441, 2017.

M. S. Ansari, V. Mrazek, B. F. Cockburn, L. Sekanina, Z. Vasicek and J. Han, "Improving the Accuracy and Hardware Efficiency of Neural Networks Using Approximate Multipliers," in *IEEE Transactions on Very Large Scale Integration (VLSI) Systems*, vol. 28, no. 2, pp. 317–328, Feb. 2020.

A. Momeni, J. Han, P. Montuschi and F. Lombardi, "Design and Analysis of Approximate Compressors for Multiplication," in *IEEE Transactions on Computers*, vol. 64, no. 4, pp. 984–994, April 2015.

R. R. Osorio and G. Rodríguez, "Truncated SIMD Multiplier Architecture for Approximate Computing in Low-Power Programmable Processors," in *IEEE Access*, vol. 7, pp. 56353-56366, 2019.

S. Morgan, R. A. Anderson, C. Gourley, W. H. Wallace, N. Spears, "How do chemotherapeutic agents damage the ovary?," *Human Reproduction Update,* vol. 18, 5, 525–535, 2012.

A. M. H. Koning, W. K. H. Kuchenbecker, H. Groen, A. Hoek, J. A. Land, K. S. Khan, B. W. J. Mol, "Economic consequences of overweight and obesity in infertility: a framework for evaluating the costs and outcomes of fertility care," *Human Reproduction Update*, vol.16, I3, 246–254, 2010.

M. R. Azghadi et al., "Hardware Implementation of Deep Network Accelerators Towards Healthcare and Biomedical Applications," in *IEEE Transactions on Biomedical Circuits and Systems*, vol. 14, 6, 1138–1159, 2020.

E. Ceolini *et al.*, "Hand-gesture recognition based on EMG and eventbased camera sensor fusion: A benchmark in neuromorphic computing," *Frontiers in Neuroscience.*, vol. 14, 520438, 637, 2020.

Y. Chen, T. Krishna, J. S. Emer and V. Sze, "Eyeriss: An Energy-Efficient Reconfigurable Accelerator for Deep Convolutional Neural Networks," in *IEEE Journal of Solid-State Circuits*, vol. 52, 1, 127–138, 2017.

Q. Guan et al., "Deep convolutional neural network VGG-16 model for differential diagnosing of papillary thyroid carcinomas in cytological images: A pilot study," *Journal of Cancer*, vol. 10, 20, 2019.

N. Tajbakhsh et al., "Convolutional neural networks for medical image analysis: full training or fine tuning?" *IEEE Transactions on Medical Imaging*, vol. 35, 5, 1299–1312, 2016.

C. Lammie, W. Xiang, B. Linares-Barranco, and M. R. Azghadi, "Mem-Torch: An open-source simulation framework for memristive deep learning systems," , arXiv:2004.10971, 2020.

A. M. Hassan, A. F. Khalaf, K. S. Sayed, H. H. Li, and Y..Chen, "Real-time cardiac arrhythmia classification using memristor neuromorphic computing system," in Proceedings of the Conference of the IEEE Engineering in Medicine and Biology Society (EMBC), pp. 2567–2570, 2018.

F. Cai et al., "A fully integrated reprogrammable memristor–CMOS system for efficient multiply–accumulate operations," *Nature Electronics*, vol. 2, no. 7, pp. 290–299, 2019.

L. Ohno-Machado and D. Bialek, "Diagnosing breast cancer from fnas: variable relevance in neural network and logistic regression models," *Student Health Technology Information*, vol. 52, pp. 537–540, 1998.

Y. Ku, W. Tompkins, and Q. Xue, "Artificial neural network for ECG arrhythmia monitoring," In proceedings of IJCNN International Joint Conference on Neural Networks, vol. 2, pp. 987–992, 1992.

S. M. McKinney et al., "International evaluation of an AI system for breast cancer screening," *Nature*, vol. 577, no. 7788, pp. 89–94, 2020.

E. Nurse, B. S. Mashford, A. J. Yepes, I. Kiral-Kornek, S. Harrer, and D. R. Freestone, "Decoding EEG and LFP signals using deep learning: Heading True North," in Proceedings of ACM International Conference on Computing Frontiers, pp. 259–266, 2016.

A. Jafari, A. Ganesan, C. S. K. Thalisetty, V. Sivasubramanian, T. Oates, and T. Mohsenin, "SensorNet: A scalable and low-power deep convolutional neural network for multimodal data classification," *IEEE Transactions on Circuits and Systems I: Regular Papers*, vol. 66, no. 1, pp. 274–287, 2019.

T. Hirtzlin et al., "Digital biologically plausible implementation of binarized neural networks with differential hafnium oxide resistive memory arrays," *Frontiers of Neuroscience*, vol. 13, pp. 1383, 2020.

F. Corradi et al., "ECG-based heartbeat classification in neuromorphic hardware," in Proceedings of International Joint Conference on Neural Networks, pp. 1–8, 2019.

E. Ceolini et al., "Hand-gesture recognition based on EMG and event based camera sensor fusion: A benchmark in neuromorphic computing," *Frontiers of Neuroscience*, vol. 14, no. 520438, 2020.

F. C. Bauer, D. R. Muir, and G. Indiveri, "Real-time ultra-low power ECG anomaly detection using an event-driven neuromorphic processor," *IEEE Transactions on Biomedical Circuits and Systems*, vol. 13, no. 6, pp. 1575–1582, 2019.

E. Donati et al., "Processing EMG signals using reservoir computing on an event-based neuromorphic system," in Proceedings of the IEEE Biomedical Circuits Systems Conference, pp. 1–4, 2018.

J. Behrenbeck et al., "Classification and regression of spatio-temporal signals using NeuCube and its realization on SpiNNaker neuromorphic hardware," *Journal of Neural Engineering*, vol. 16, no. 2, pp. 026014, 2019.

M. Wess, P. S. Manoj, and A. Jantsch, "Neural network based ECG anomaly detection on FPGA and trade-off analysis," In Proceedings of IEEE International Symposium on Circuits and Systems, pp. 1–4, 2017.

F. Xing, Y. Xie, X. Shi, P. Chen, Z. Zhang, and L. Yang, "Towards pixel to-pixel deep nucleus detection in microscopy images," *BMC Bioinformatics*, vol. 20, no. 1, pp. 1–16, 2019.

R. R. Shrivastwa, V. Pudi, and A. Chattopadhyay, "An FPGA-based brain computer interfacing using compressive sensing and machine learning," in Proceedings of IEEE Computer Society Annual Symposium VLSI, pp. 726–731, 2018.

P. Škoda, T. Lipi´c, Srp, B. M. Rogina, K. Skala, and F. Vajda, "Implementation framework for artificial neural networks on FPGA," In Proceedings of the 34th International Convention MIPRO, pp. 274–278, 2011.

N. Ibrahim, S. Salah, M. Safar and M. W. El-Kharashi, "Digital Design using CMOS and Hybrid CMOS/Memristor Gates: A Comparative Study," 2018 13th International Conference on Computer Engineering and Systems (ICCES), 2018

A. M. Sodagar, "Implantable Biomedical Microsystems: A New Graduate Course in Biomedical Circuits and Systems," in *IEEE Transactions on Education*, vol. 57, no. 1, pp. 48–53, 2014.

Z. Aizaz and K. Khare, "Area and Power efficient Truncated Booth Multipliers using Approximate Carry based Error Compensation," in *IEEE Transactions on Circuits and Systems II: Express Briefs*, 2021, doi: 10.1109/TCSII.2021.3094910.

R. J. Martis et al., "Application of principal component analysis to ECG signals for automated diagnosis of cardiac health," *Expert Systems Application*, vol. 39, no. 14, pp. 11792–11800, 2012.

A. H. Khandoker, M. Palaniswami, and C. K. Karmakar, "Support vector machines for automated recognition of obstructive sleep apnea syndrome from ECG recordings," in *IEEE Transactions on Information Technology in Biomedicine*, vol. 13, no. 1, pp. 37–48, 2009.

G. Li and W.-Y. Chung, "Detection of driver drowsiness using wavelet analysis of heart rate variability and a support vector machine classifier, *Sensors*, vol. 13, no. 12, pp. 16494–16511, 2013.

Q. Li, C. Rajagopalan, and G. D. Clifford, "Ventricular fibrillation and tachycardia classification using a machine learning approach," *IEEE Transactions on Biomedical Engineering*, vol. 61, no. 6, pp. 1607–1613, Jun. 2014.

M. Bsoul, H. Minn and L. Tamil, "Apnea MedAssist: Real-time Sleep Apnea Monitor Using Single-Lead ECG," *IEEE Transactions on Information Technology in Biomedicine,* vol. 15, no. 3, pp. 416–427, 2011.

X.-W. Wang, D. Nie, and B.-L. Lu, "EEG-based emotion recognition using frequency domain features and support vector machines," in Proceedings of 18th ICONIP Conference, Shanghai, China, pp. 734–743, 2011

X. W. Wang, D. Nie, and B. L. Lu, "Emotional state classification from EEG data using machine learning approach," *Neurocomputing*, vol. 129, no. 4, pp. 94–106, 2014.

R. Panda, P. S. Khobragade, P. D. Jambhule, S. N. Jengthe, P. R. Pal, and T. K. Gandhi, "Classification of EEG signal using wavelet transform and support vector machine for epileptic seizure prediction," in Proceedings of International Conference on Systems in Medicine and Biology (ICSMB), Kharagpur, India, pp. 405–408, 2010.

Y. Liu, W. Zhou, Q. Yuan, and S. Chen, "Automatic Seizure Detection Using Wavelet Transform and SVM in Long-Term Intracranial EEG," in *IEEE Transactions on Neural Systems and Rehabilitation Engineering*, vol. 20, no. 6, pp. 749–755, 2012.

U. R. Acharya et al., (a)"Automated detection of arrhythmias using different intervals of tachycardia ECG segments with convolutional neural network," *Information Sciences*, vol. 405, pp. 81–90, 2017.

U. R. Acharya et al., (b) "Automated detection of coronary artery disease using different durations of ECG segments with convolutional neural network," *Knowledge Based Systems*, vol. 132, pp. 62–71, 2017.

U. R. Acharya et al., (c) "Automated identification of shockable and non shockable life-threatening ventricular arrhythmias using convolutional neural network," *Future Generation Computer Systems*, vol. 79, pp. 952–959, 2018.

U. R. Acharya et al., (d) "Application of deep convolutional neural network for automated detection ofmyocardial infarction using ECG signals," *Information Sciences*, vol. 415, pp. 190– 198, 2017.

U. R. Acharya et al., "A deep convolutional neural network model to classify heartbeats," *Computers in Biology and Medicine*, vol. 89, pp. 389–396, 2017.

W. Zheng and B. Lu, "Investigating critical frequency bands and channels for EEG-based emotion recognition with deep neural networks," *IEEE Transactions on Autonomous Mental Development*, vol. 7, no. 3, pp. 162–175, 2015.

J. Suwicha, P. N. Setha, and I. Pasin, "EEG-based emotion recognition using deep learning network with principal component-based covariate shift adaptation," *Scientific World*, vol. 2014, pp. 1–10, 2014.

E. S. Pane, A. D..Wibawa, and M. H. Purnomo, "Improving the accuracy of EEG emotion recognition by combining valence lateralization and ensemble learning with tuning parameters," *Cognitive Processing*, pp. 1–13, 2019.

U. E. Derya, "Combining recurrent neural networks with eigenvector methods for classification of ECG beats," *Digital Signal Processing*, vol. 19, no. 2, pp. 302–329, 2009.

U. E. Derya, "Recurrent neural networks employing Lyapunov exponents for analysis of ECG signals," *Expert Systems with Applications*, vol. 37, no. 2, pp. 1192–1199, 2010.

A. T. Tzallas, M. G. Tsipouras, and D. I. Fotiadis, "Epileptic seizure detection in eegs using time–frequency analysis," *IEEE Transactions on Information Technology in Biomedicine*, vol. 13, no. 5, pp. 703–710, Sep. 2009.

S. M. S. Alam and M. I. H. Bhuiyan, "Detection of seizure and epilepsy using higher order statistics in the EMD domain," *IEEE Journal on Biomedical Health Information*, vol. 17, no. 2, pp. 312–318, 2013.

J., Yang, D., Strukov & D., Stewart, "Memristive devices for computing". *Nature Nanotechnology*, vol. 8, pp. 13–24, 2013

11 Cross-Domain Analysis of Social Data and the Effect of Valence Shifters

Aarti
Department of Computer Science and Engineering, Lovely Professional University

Raju Pal
Department of Computer Science and Engineering, Jaypee Institute of Information Technology, Noida, India

Contents

11.1	Introduction	228
	11.1.1 Machine Learning	229
	11.1.2 In a Lexical-Based Approach We Have Two Basic Approaches	230
11.2	Literature Survey	230
	11.2.1 Sentiment Analysis Packages and Techniques	231
	11.2.2 Data Extraction for Sentimental Analysis	232
	11.2.3 Sentiment Analysis Survey, Classifiers, and N Grams	232
	11.2.4 Sentiment Analysis Using Big Data Tools	232
	11.2.5 Sentimental Analysis Using Websites	233
	11.2.6 Sentiment Analysis of Social Sites	233
11.3	Prerequisite Knowledge/Technique Used	233

DOI: 10.1201/9781003189633-11

	11.3.1	Social Media	233
	11.3.2	Sentimental Analysis	234
	11.3.3	Social Sentiment Analysis	235
	11.3.4	Naïve-Bayesian	235
	11.3.5	N Grams	235
	11.3.6	Data Set	235
	11.3.7	Valance Shifters	236
	11.3.8	Confusion Matrix	236
11.4	Methodology		236
	11.4.1	Data Extraction	237
	11.4.2	Removing Duplicity	237
	11.4.3	Creating Corpus	237
	11.4.4	Text Cleaning	238
	11.4.5	Stopwords Removal	238
	11.4.6	N Gram Generation	238
	11.4.7	Sentiment Analysis	238
11.5	Result and Discussion		238
11.6	Conclusion and Future Scope		243
References			244

11.1 INTRODUCTION

Sentiment analysis is the process of understanding the sentiments from text, which defines the text to be positive, negative, or neutral. It is easy for humans to understand the opinions from text by simply reading it but for computers we need some specialized techniques. With increasing amounts of data over the internet, we need such specialized techniques that are capable of understanding the sentiments from text. Sentiment analysis can be turned into a question of whether a piece of text is expressing positive, negative, or neutral sentiment towards the discussed topic and can be thus understand as a knowledge-based classification problem. Nowadays, the undesirable truth is that social media users are more likely to share now how they feel about a current "hot topic" on social media platforms. Users may post positive or neutral opinions about that topic or a particular products they are using. Sentiment analysis, also known as opinion mining, is a process of analyzing the polarity of the emotion behind a line of text. In the social media that contains a huge amount of texts and a large range of topics, it would be very difficult to manually collect enough labeled data to train a sentiment classifier for different domains. Distant supervision that considers emoticons as natural sentiment labels in the microblogs texts has been widely used in social media sentiment analysis. To address such challenges, in this work we propose a supervised lifelong learning framework for large-scale social media sentiment analysis. We can use different machine learning and lexical analysis techniques for such purposes.

11.1.1 Machine Learning

Using machine learning algorithms that can receive input data, we can use statistical analysis to predict output value in a desired range. In machine learning, we mostly use supervised learning techniques for sentimental analysis. The following types of classifiers can be used for supervised learning for sentimental analysis:

- *Decision tree classifiers*: In a decision tree classifier, tree structure is followed. The scanning starts with the root node and then moves to the internal or leaf node. At the internal node, the condition is checked at that node. After reaching the leaf node, an output or decision can be made. In a decision tree, direct movement to the internal node cannot be done without solving the root node.
- *Linear classifiers*: Linear classifiers are those that use the results from the linear classification results to make a prediction. In this, two kinds of classifiers are used: (i) in support vector machines, a hyperplane is used that best divides the data set into two classes. This algorithm solves the problem by dividing it into sub-problems that are solved analytically. As the linear equality constraint involves the Lagrange multipliers a_i, the smallest possible problem involves two such multipliers and $y_i = \in \{-1, +1\}$ is a binary output label. Then, for any two multipliers a_1 and a_2, the constraints are reduced to:

$$0 \leq a_1 a_2 \leq C$$
$$y_1 a_1 + y_2 a_2 = k$$

This reduced problem can be solved analytically. In the equation, C is a SVM hyperparameter and K is the negative of the sum divided by the rest of terms in the equality constraint, which is fixed on each iteration.

- **Neural networks** are based on the neurons of brains and are used for deep learning techniques. ANN consists of processing units that communicate by sending signals with the help of weighted connections to each other. ANN consists of processing units (neurons), state of activation, the connection between processing units (weights), an activation function, an external input (bias), and learning rules. ANN can be defined with a triplet (N, C, W) where N is set of neurons, C is set of $\{(i, j) | i, j, \in N\}$ that determines the connection between neuron i and neuron j, and w is weight matrix w (i, j) that is weight between neuron i and j. Suppose propagation function (n_j) of j receive n number of outputs $o_1, o_2, o_3 \ldots O_n$ of neurons i_1, i_2, i_3 that transfers connection weight between them to network input, which is processed by activation function.
- **Probabilistic classifiers:** These classifiers have an additional advantage that they are capable of predicting the input as well. These classifiers are derived

from generative probability models, which provide a principled way to the study of statistical classification in complex domains such as natural language and visual processing. Some of probabilistic classifiers are Naïve-Bayes, Bayesian network, and Maximum entropy.

11.1.2 In a Lexical-Based Approach We Have Two Basic Approaches

- **Dictionary-based approach:** In a dictionary-based approach, dictionaries defining words on a lexical scale are used. In a lexical scale, the magnitude of positivity or the magnitude of negativity of a word is defined.
- **Corpus-based approach:** In corpus-based techniques, preprocessing and cleaning of data and corpus creation of that data takes place.

We can also distinguish sentiment analysis techniques as different levels of persistence. There are three different levels for sentiment analysis.

- **Document level:** In document-level sentiment analysis, we consider the whole document to be a single unit and define the whole document to be positive, negative, or neutral. This type of sentiment analysis does not perform for a multi-topic textual unit.
- **Sentence level:** In this level, we consider a single sentence to be the unit and we define the sentences to be positive, negative, or neutral. At this level, sentiment analysis can be performed for multi-topic text as well. Each sentence is taken as a unit and then the level of positivity or negativity is checked in that sentence. This magnitude of positivity or negativity is referred to as polarity.
- **Entity or aspect level:** It the deepest persisting level of analysis, a single word or a phrase is considered under analysis and defines the entities to be positive, negative, and neutral. At this level of sentiment analysis, the working is done on the smallest units.

11.2 LITERATURE SURVEY

Sentiment analysis was not a very popular topic two decades back. There was not much data present online before that; therefore, considering automation for sentiment analysis was not considered. With the growth of data on the internet after the invention of social networking sites, the data over the internet increased. Many people presented their views on the internet in an unbiased way and thus this data was considered of importance for knowledge retrieval. But on the other hand, there were also some challenges involved. The first one was that this data was not clean data. Secondly, it is difficult for machines to understand the sentiments from text language like humans. Thirdly, the data is present in large volumes; therefore, it is difficult for organizations to appoint humans for sentiment analysis.

For these challenges, researchers have worked in this field. They have applied different data preprocessing techniques for cleaning the data, such as NLP (natural language processing) systems that can be used to extend the act as language understanding systems to understand sentiments from text data.

There were not much papers published on this topic before the invention of social sites because most of the data that is present is through social media and microblogging sites. Therefore, most of the data was generated after that.

We have divided the literature survey under different topics. Section (11.2.1) contains the papers for the basic sentimental analysis and few techniques for sentiment analysis. Section (11.2.2) contains the papers for data extraction from social media that can later be used for sentimental analysis. Section (11.2.3) contains the survey paper on sentimental analysis and the classifiers and other important elements that can be useful in increasing the accuracy of sentimental analysis. Section (11.2.4) contains the papers of sentimental analysis using big data tools. Section (11.2.5) contains the literature regarding performing the sentimental analysis by extracting data from websites. Section (11.2.6) contains the papers of sentimental analysis of the social data. Lastly, Section (11.2.7) contains the comparison table for literature.

11.2.1 Sentiment Analysis Packages and Techniques

Although this topic has been partially discussed in previous literature, sentiment analyzer is literature that was truly on this topic was published in the year 2003 [1]. In this paper, author did not considered the whole topic but performed sentiment analysis at feature level. A complete study was done on sentiment analysis. The reviews from different domains, especially from movies reviews and camera reviews were taken as a data set and sentiments were analyzed. After this, the study on sentiment analysis became more frequent as social data increased and people like to read reviews before buying any products. Different strategies were followed after this to increase the efficiency in analysis of data.

In the year 2009, Lipika [2] worked on noisy data for sentiment analysis. They created a new WordNet package for noisy data. This was a good step because the data that we get from social or microblogging websites is noisy data. Therefore, treating the noisy data technique is helpful when it comes to sentiment analysis from social or microblogging websites.

In the year 2011, Jingbo [3] worked on some restaurant data that consider single data and multi-aspect data techniques to help the restaurant owner in understanding views of customers. Chenghua [4] in 2011 considered the JST technique to classify data and was able to create a weakly supervised technique that was capable of working with higher rate of efficiency by being trained on small data. Leung [5] created a probalostic rating interface to rate the text on a numerical scale. After this, some authors worked on Wordnet (famous sentiment analysis package). Alena Neviarouskaya [6] and Danushka Bollegala [7] tried to increase the efficiency of sentiment analysis by considering synonyms and antonynms in addition to words of WorNet. Further, Uros Krcadinac [8] considered

Ekman emotion classification using WordNet. He used two parts for sentiment analysis: word lexicon and emotion lexicon, which makes the technique more suitable for analysis in social media.

In 2015, Lorenzo Gatti [9] created SentiWords in which he created prior polarity lexica and we can have high precision and high coverage. Abinash Tripathy [10] considered a hybrid appriach for sentiment analysis. He considered artificial neural networks and data mining classification algorithms for sentiment analysis. A layer of neural networks was used for improving the results. Evaluation of the results was done using precison, recall, f-measure, and accuracy.

11.2.2 Data Extraction for Sentimental Analysis

For extraction of data, most commonly used techniques are crawling data from websites. But in the case of social networking sites, they provide access to data using an application program interface. Howden Chris [11] used OAuth twitter application for data extraction and used it for forensic purposes. Similarly, in 2015, Rehab [12] created RUM Extractor, which was based on the similar facebook API. He used Php as programming language and Apache server to create a software for easy access to social data.

11.2.3 Sentiment Analysis Survey, Classifiers, and N Grams

In 2008, Bo Pang [13] discussed the concept on opinion and sentiment analysis in detail, concentrating on difficulties and challenges in this field. In 2010, Hsinchun Chen [14] also discussed the past implementations and techniques. He concluded that high perfection levels in sentiment analysis are not reached yet because enough work has not been done on this topic. Anuj Sharma [15] compared different machine learning and feature extraction techniques and came to the conclusion that SVM performs best in machine learning and Naïve-Bayesian as the best classifier. Dhiraj [16] used a feature vector of test data to create an automated program for sentiment analysis of social data. In the year 2014, a survey was again done in sentiment analysis. Walla [17] made a survey from very initial level and concluded that Naïve-Bayesian and SVM are the most frequently used in sentiment analysis. One of the major challenges in sentiment analysis are that most of the techniques do not work on multiple languages. Fotis Asiopos [18] tried to make a technique that could work for multiple languages on social data.

11.2.4 Sentiment Analysis Using Big Data Tools

Big data tools have emerged as a savior technique to make an analysis from a large volume of data in less time compared to simple tools. Some authors have tried to make the sentiment analysis task fast using the big data tools. Sudipto [19] tried to make analysis fast using big data tools and was able to achieve accuracy of 67.6%. Devendra K Tayal [20] made a try using a Bloom filter with Hadop for faster processing.

11.2.5 Sentimental Analysis Using Websites

Many authors tried to make use of data from websites to make a sentiment analysis and then benefit the organizations in business. L. Zhang [21] made a business intelligence system by making sentiment analysis from data crawled from websites. Shoiab Ahmed [22,23] published two papers in 2015. In the first paper, he simply worked on sentiment analysis using different rule-based classifiers and in the second, he implemented the classifier using POS (parts of speech) tagging.

11.2.6 Sentiment Analysis of Social Sites

These days people express their view freely on social networking sites. We can find unbiased views about some particular topic on social media websites. People prefer to read reviews on social networking sites before buying a product. Therefore, sentiment analysis from social media can help customers as well as organizations to have a clear feedback about their products.

For these reasons, many authors worked on sentiment analysis from social media. Rahim Dehkharghani [24] used causal rules on social data for analysis. Alexandra Balahur [25] used SVM on multi-language tweets and found that n gram improves the efficiency of such a system. Different techniques have been used for overcoming different challenges in sentiment analysis. Matthijs Meire [26] made use of the pre-post and after-post data for making the analysis results more efficient. He considered the personal information and history of the person and how he has reacted to different domain posts. Cambria [27] moved a step forward and explained how social media is not just limited to building a social network but it could also be used for business intelligence or corporate purposes.

Initially, we had two main types of techniques for sentiment analysis. One is machine learning and another one is lexical analysis. But with time for making the analysis more accurate, authors have adopted hybrid approaches, some of which they were able to get more efficient results. Avinash Chandra Pandey [28] also used a hybrid approach of evolutionary cuckoo search and k-means for sentiment analysis. He also considered emotion symbols and was able to give good results in terms of accuracy.

11.3 PREREQUISITE KNOWLEDGE/ TECHNIQUE USED

11.3.1 Social Media

Social media is the data present over the internet. The data may be from any websites, forums, social networking sites, microblogging sites, etc. Earlier this data was small and least consideration was given on analysis of data present online. The data that was considered for analysis was only the basic data. But with the advent of social media and micro-blogging websites, the data on the internet increased a lot and each part of this data is considered for analysis purposes.

Slowly and gradually this data became commercialized. Initially social media was considered useful for learning information but as time passed, the social media and commercial world became very close and the data is popularly known as social commercial data.

These days the social media is not just used as a medium for social connectedness or learning purposes, but it has also become commercialized. Large organizations have built their social pages where the users express their views freely in an unbiased way. When they analyzed the data from social media and made decisions, it came out to be of great benefit. These days, social media marketing and analysis are hot topics and many new techniques are being used to make the procedure more efficient.

11.3.2 Sentimental Analysis

Sentiment analysis is the process of understanding the sentiment of people from a piece of text. It is not a very difficult task for humans to interpret to make out that what a person is trying to convey from text written. With the increase in amount of data, it became necessary that the task of analyzing sentiments must be automated. This task will make the procedure easier and large samples of text can be analyzed in far less time. But it is a very challenging task to create such a system that can understand the sentiments like humans.

Researchers have tried to make such automated systems and are able to create systems that can review sentiments with high efficiency but these are not applicable to cross-domain data.

There are three basic levels of sentiment analysis.

- Document-level sentiment analysis: In document-level sentiment analysis, we consider the whole document to be a single unit and single topic. We can consider single topic documents in this and it can be considered in the case of analyzing topic from text.
- Sentence-level sentiment analysis: In sentence-level sentiment analysis, we consider a sentence to be a single unit. It is capable of analyzing multiple topic documents. In this case, we define the subjectivity at sentence level i.e. we define that whether a sentence is positive or negative. Based on the opinions on all of the sentences, we can make a final decision of opinion for the whole document.
- Word or entity-level sentiment analysis: At the entity or feature level, we calculate the opinion at the feature or word level. This is a much more detailed procedure to calculate the sentiments. At this level, we can make out each and every detail of opinion mining. Thus, this is considered to be the most detailed and efficient way of sentiment analysis. Therefore, it is also the most practiced technique for sentiment analysis.

11.3.3 Social Sentiment Analysis

Social sentiment analysis is the sentiment analysis done on social data. Social sentiment analysis is the basic way of extracting knowledge from social data. Further, this knowledge extracted from social data can be used for various commercial purposes. There are many challenges in social data sentiment analysis like it being used for cleaning data, feature extraction, and the most important one is to perform the sentiment analysis at a high efficiency rate.

11.3.4 Naïve-Bayesian

Naïve-Bayesian is a classification algorithm based on the Bayes theorem. The Bayes theorem describes the probability of an event, based on prior knowledge of conditions that might be related to the event. In Naïve-Bayesian, no relatedness is considered between two features that is why it is called naïve. This algorithm is considered best for text classification and is capable of giving tough competition to much more complex methods such as support vector machines. In Naïve-Bayesian, we calculate posterior probability. It can be simply defined with the following equation:

$$P\frac{C}{X} = \frac{P\frac{X}{(C)}P(C)}{p(X)}$$

P(C/X) is the posterior probability of class (c, target) given predictor (x, attributes).
P(C) is the prior probability of class.
P(C/X) is the likelihood which is the probability of predictor given class.
P(X) is the prior probability of predictor.

11.3.5 N Grams

N grams are widely used in statistical natural language processing systems. These are widely used in speech recognition. The basic idea of an n gram is that a consecutive sequential stream of words are used. For example, let us consider a sentence, "He is an honest person." Then the words can be considered as grams and we can consider any number of words as grams according to our preference. For instance, if we consider one word at a time then it is a unigram; if we consider two words it is bigram; and if we consider three consecutive words then it is called a trigram and so on.

The following example clearly explains the n gram. It explains how we can divide words in different scenarios (Table 11.1).

11.3.6 Data Set

We have used the data set created by Blitzer [29] for multi-domain data. The author has collected and categorized the data of four different domains.

TABLE 11.1 Example of N gram

SENTENCE	UNIGRAM	BIGRAM	TRIGRAM
He is an honest person.	honest, person.	He is, is an, an honest, honest person.	He is an, is an honest, an honest person.

11.3.7 Valance Shifters

In sentimental analysis, we are supposed to work on text data and we have to consider various facts and events. But we can also consider the attitude of the writer to see their results on the accuracy of the sentiment analysis. For considering the attitude of the writer, we consider valence shifters. In valence shifters, the words that are more emphasized are given more magnitude in the calculation of sentimental analysis.

11.3.8 Confusion Matrix

Confusion matrix is a technique that can be used in the sentiment analysis for process of accuracy checking. Here we consider four main categories:

- **True Positive (TP):** This is the condition in which the actual value is positive and we also consider the value to be positive.
- **True Negative (TN):** This is the condition where the actual value is negative and we also consider the value to be negative.
- **False Positive (FP):** This is the condition where the actual value is negative but we consider the value to be positive.
- **False Negative (FN):** This is the condition where the actual value is positive and we consider the value to be negative.

Based on these conditions, we have formulas where we can find the accuracy, precision, and recall. There are also some other terms like specificity and degree of positivity that could also be considered for an efficiency check.

11.4 METHODOLOGY

From the literature survey, we were able to find that the main challenges in sentiment analysis are

- Multi-domain data was not processed efficiently in any of the techniques.
- Multi-language data was also considered but it worked only in domain-specific trained data.

We are going to build a technique for our first challenge. We are going to use N grams in combination with a hybrid approach. N grams are considered good in sentiment analysis because they are a sequence of words that we can consider as a word multiple times instead of considering it only once in the analysis. For example, when we are using sentiment analysis using bigrams we consider each word twice in analysis.

The other main challenges of doing sentiment analysis on social data are preprocessing of data and making the data ready for analysis purposes by applying various data cleaning techniques. Following is the methodology that we followed for sentiment analysis in a cross domain.

11.4.1 Data Extraction

Various methods can be applied for extraction of data in a sentiment analysis process. In our case, we are working on social data. Some are extracting data from social networking or microblogging websites. The most common and easy way of extracting data from the social networking websites is using the application programming interface provided by the social networking sites. They provide different penetration levels to data and different privileges according to the service demanded by the user. They provide a deep penetration level if we are using OAuth application and getting users permission to access data; else, we have less permissions.

Crawling data from the websites is another method that can be used for data extraction. It is the other most commonly used method for data extraction.

11.4.2 Removing Duplicity

After crawling data, we have the initial step of removing the duplicate records. Many times we find that the reviews are submitted twice by a user. These are removed to save time and increase efficiency of the analysis. There can also be noisy data; sometimes we arrive at a condition that some advertisers are posting advertisements in the reviews for their publicity. A common phenomenon seen is that an advertiser advertises multiple times in the reviews. So, this duplication removal may also help in removal of those unwanted advertisements.

11.4.3 Creating Corpus

Once we get the distinct records, our next step is to continue with other noise removing and data cleaning operations. These operations can be performed more easily on corpus. So, we create a corpus of the review and continue with the text cleaning operations. A corpus is also beneficial in a way that processes can be easily applied to a corpus.

11.4.4 Text Cleaning

The data that we get from the social networking sites is not noise free and clean data. So, before proceeding with further processing of the data, we must clean the data. First of all, we must check duplicate records and remove the duplicity from the records. Then we can remove punctuations, other unwanted symbols, and URLs. According to a situation, we can also customize the solution for removing the numbers as well. These numbers are mostly mobile numbers entered by the reviewer.

11.4.5 Stopwords Removal

After removing duplicity and cleaning, we can continue with the next step of removing stopwords from the data. Stopwords are those words that are frequently used in language but have least significance in analysis. Words such as is, am, are, this, etc. are removed as they are considered to be least significant in sentiment analysis.

11.4.6 N Gram Generation

After cleaning the data, the n grams of that data are created. The fact is considered that a word taken more than once for polarity checking provides more efficient results. N grams are one in which we consider the sequence of words and in the next step the sentiment analysis is done for sequence of words. By doing this we may consider a word multiple times, which increases the efficiency of the solution.

11.4.7 Sentiment Analysis

The n grams are provided as input to the sentiment analyzer. In sentiment analysis, the analyzer provides polarity of different words. Polarity is the magnitude of positivity or negativity of the words or the phrases that are considered. Here, at this step, we consider two cases: one considering the valance shifters and the other without considering the valence shifters. Figure 11.1 describes the whole scenario of proposed methodology. The graphical representation of all the steps from 4.1 to 4.7 is depicted in Figure 11.1.

11.5 RESULT AND DISCUSSION

After analysis of data, results are evaluated considering the effect of valence shifters. The accuracy of the analysis considering valence shifters is more compared to the one in which we do not consider the valence shifters. We have used confusion matrix for

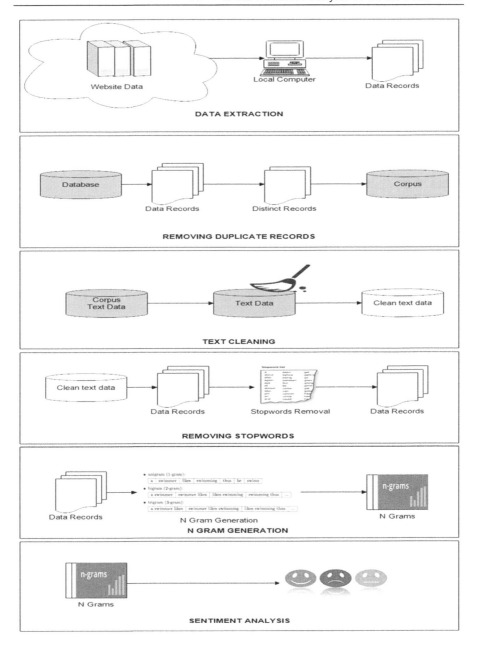

FIGURE 11.1 Steps of methodology.

checking the accuracy. We used a multi-domain data set created by Blizerd for checking the accuracy of the algorithm. Blizerd collected data from four domains, namely books, DVDs, electronics, and kitchens. In each domain, he collected 2,000 reviews from which 1,000 are positive and 1,000 are negative (Tables 11.2 and 11.3).

TABLE 11.2 Confusion matrix without considering valence shifter

DOMAIN	TN	TP	FN	FP
Books	372	771	229	628
DVDs	414	774	226	586
Electronics	313	753	247	687
Kitchen	369	762	238	631

TABLE 11.3 Confusion matrix considering valence shifter

DOMAIN	TN	TP	FN	FP
Books	362	861	139	638
DVDs	417	869	131	583
Electronics	376	904	96	624
Kitchen	360	922	78	640

Considering the values from these, we have created an accuracy, precision, and recall graph for both cases. Accuracy is defined as the total number of positive as well as negative predictions that are found to be correct.

$$\text{Accuracy} = \frac{TN + TP}{TP + TN + FP + FN}$$

Precision: It is calculated as the proportion of predictive cases that are correct. It is calculated as:

$$\text{Precision} = \frac{TP}{FP + TP}$$

Probability of detection (pd): pd is also named as recall or true-positive rate (TPR) that defines the number of positive classes that are found to be correct. It is defined by the following equation:

$$\text{Probability of detection} = \frac{TP}{TP + FN}$$

Considering the values from these, we have created an accuracy, precision, and recall graph for both cases.

Figures 11.2, 11.3, and 11.4 depict the improvement in the accuracy, precision, and recall values using valence shifters. Furthermore, the prediction model with a higher precision value shows that estimates are more accurate than others. The prediction results are compared with and without valence shifters. The proposed model

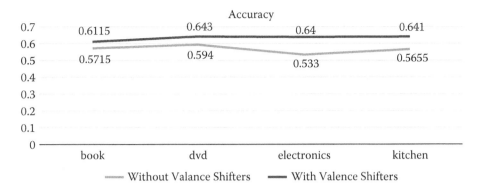

FIGURE 11.2 Accuracy comparison graph using valence shifters and without using valence shifters.

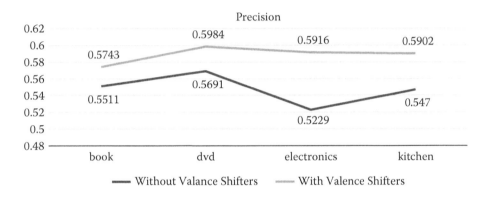

FIGURE 11.3 Comparison graph of precision using valence shifters and without using valence shifters.

FIGURE 11.4 Comparison graph of recall using valence shifters and without using valence shifters.

resulted in persistent estimates in precision, recall, and accuracy values. The following observations are concluded from previous figures:

- The capability of the model depends upon the accuracy and precision of the model. A valence shifter based model shows the appreciable and proportional results than without valence shifters.
- The predictive quality of the model expanded if the number of classes is correctly classified as faulty and the pf value is low. The recall value gives comparable and analogous results for valence shifters.

Table 11.4 describes the comparative table of proposed methodology and other techniques. The other authors do not use the concept of valence shifters. But, with the use of valance shifters, accuracy of the model is enhanced.

TABLE 11.4 Comparative analysis with other methods

SR. NO	ARTICLE NAME	AUTHOR NAME/ JOURNAL/ VOLUME NO./ PAGE NO./ YEAR	RESULTS/ FINDINGS	CONCLUSION
1	Geospatial Event Detection by Grouping Emotion Contagion in Social Media	Brendon et al. IEEE/volume-6/ pp. 159–170/ 2020	The findings suggest that Robert Parks' theory on Expressive Groups and Gustave Le Bons' theory on Social Contagion hold true in the Twittersphere.	Using graph theory and topic modelling were able to match words, hashtags, and people that are popular with the time that cities are extreme, in order to conclude what events or topics are causing this spike in emotion.
23	CAPER: Context-Aware Personalized Emoji Recommendation	Guoshuai et al. IEEE/pp. 1- 13/ 2020	Researcher conducts a series of experiments on the real-world data sets, and experiment results show better performance	Researcher proposed a context- aware personalized emojis recommendation model by considering the contextual and personal information
3	Hybrid context enriched deep learning model for fine-grained	Akshi et al. ELSEVIER/ 2020	The accuracy achieved by the proposed model is nearly 91%,	As an important finding, the decision module could also help to

TABLE 11.4 (Continued) Comparative analysis with other methods

SR. NO	ARTICLE NAME	AUTHOR NAME/ JOURNAL/ VOLUME NO./ PAGE NO./ YEAR	RESULTS/ FINDINGS	CONCLUSION
	sentiment analysis in textual and visual semiotic modality social data		which is an improvement over the accuracy obtained by the text and image modules individuals.	identify cases of neutral ambiguities, which were representative of sarcasm.
4	Adapting Deep Learning for Sentiment Classification of Code-Switched Informal Short Text	Asim et al. ACM/ 2020	The result shows that the proposed model performs better in general and adapting character-based embeddings yield equivalent performance	The proposed model McM achieves highest F1-score. Among the rest of the models, ConvNet marginally outperforms Attention-LSTM.
5	Arabic Sentiment Analysis: A Systematic Literature Review	Abdullatif et al. Applied Computational Intelligence and Soft Computing/ 2019-20	The authors declare that they have no conflict of interest.	The findings of the review propose a taxonomy for sentiment classification methods. The limitation of existing approaches are highlighted in the preprocessing step, feature generation, and sentiment classification methods.

11.6 CONCLUSION AND FUTURE SCOPE

After performing the tasks, we reach a conclusion that cross-domain valence shifters predict better results. But there is still a need to work on cross-domain algorithms to increase the accuracy of the results. For further studies, we can consider the effect of emotions on analysis of data. Emoticons have been used by some of the authors in previous studies and they proved to be useful in analysis with a higher rate of efficiency. In this paper, it is performed that an emoji-based sentiment analysis and classifier to gain insights into users' emotional reactions to posts on social websites with the utilization of the Google Cloud Prediction API to estimate the sentiment of the posts from posted texts and emoji-based reactions. Due to the lack of large-scale data sets, the prevailing approach in visual sentiment analysis is to leverage models trained for object classification in large data sets like ImageNet. So, there is a need to develop a technique for visual sentimental analysis. In future work, we tried to develop the technique for the visual sentimental analysis [30–51].

REFERENCES

[1] Yi, J., Nasukawa, T., Dunescu, R., & Niblack, W. (2003, November). Sentiment analyzer: Extracting sentiments about a given topic using natural language processing techniques. In Data Mining, 2003. ICDM 2003. Third IEEE International Conference on (pp. 427–434). IEEE.

[2] Dey, L., & Haque, S. M. (2009). Opinion mining from noisy text data. *International Journal on Document Analysis and Recognition (IJDAR)*, 12(3), 205–226.

[3] Zhu, J., Wang, H., Zhu, M., Tsou, B. K., & Ma, M. (2011). Aspect-based opinion polling from customer reviews. *IEEE Transactions on Affective Computing*, 2(1), 37–49.

[4] Lin, C., He, Y., Everson, R., & Ruger, S. (2012). Weakly supervised joint sentiment-topic detection from text. *IEEE Transactions on Knowledge and Data engineering*, 24(6), 1134–1145.

[5] Leung, C. W. K., Chan, S. C. F., Chung, F. L., & Ngai, G. (2011). A probabilistic rating inference framework for mining user preferences from reviews. *World Wide Web*, 14(2), 187–215.

[6] Neviarouskaya, A., Prendinger, H., & Ishizuka, M. (2011). SentiFul: A lexicon for sentiment analysis. *IEEE Transactions on Affective Computing*, 2(1), 22–36.

[7] Bollegala, D., Weir, D., & Carroll, J. (2013). Cross-domain sentiment classification using a sentiment sensitive thesaurus. *IEEE transactions on knowledge and data engineering*, 25(8), 1719–1731.

[8] Krcadinac, U., Pasquier, P., Jovanovic, J., & Devedzic, V. (2013). Synesketch: An open source library for sentence-based emotion recognition. *IEEE Transactions on Affective Computing*, 4(3), 312–325.

[9] Gatti, L., Guerini, M., & Turchi, M. (2016). Sentiwords: Deriving a high precision and high coverage lexicon for sentiment analysis. *IEEE Transactions on Affective Computing*, 7(4), 409–421.

[10] Tripathy, A., Anand, A., & Rath, S. K. (2017). Document-level sentiment classification using hybrid machine learning approach. *Knowledge and Information Systems*, vol. 53, pp. 805–831.

[11] Howden, C., Liu, L., Li, Z., Li, J., & Antonopoulos, N. (2014). Virtual vignettes: the acquisition, analysis, and presentation of social network data. *Science China Information Sciences*, 57(3), 1–20.

[12] Duwairi, R. M., & Alfaqeeh, M. (2015, August). RUM Extractor: A Facebook Extractor for Data Analysis. In Future Internet of Things and Cloud (FiCloud), 2015 3rd International Conference on (pp. 709–713). IEEE.

[13] Pang, B., & Lee, L. (2008). Opinion mining and sentiment analysis. *Foundations and Trends® in Information Retrieval*, 2(1–2), 1–135.
[14] Chen, H., & Zimbra, D. (2010). AI and opinion mining. *IEEE Intelligent Systems*, 25(3), 74–80.
[15] Sharma, A., & Dey, S. (2012, October). A comparative study of feature selection and machine learning techniques for sentiment analysis. In Proceedings of the 2012 ACM research in applied computation symposium (pp. 1–7). ACM.
[16] Gurkhe, D., & Bhatia, R. (2014). Effective Sentiment Analysis of Social Media Datasets using Naive Bayesian Classification.
[17] Medhat, W., Hassan, A., & Korashy, H. (2014). Sentiment analysis algorithms and applications: A survey. *Ain Shams Engineering Journal*, 5(4), 1093–1113.
[18] Aisopos, F., Tzannetos, D., Violos, J., & Varvarigou, T. (2016, March). Using n-gram graphs for sentiment analysis: an extended study on Twitter. In Big Data Computing Service and Applications (BigDataService), 2016 IEEE Second International Conference on (pp. 44–51). IEEE.
[19] Dasgupta, S. S., Natarajan, S., Kaipa, K. K., Bhattacherjee, S. K., & Viswanathan, A. (2015, October). Sentiment analysis of Facebook data using Hadoop based open source technologies. In Data Science and Advanced Analytics (DSAA), 2015. 36678 2015. IEEE International Conference on (pp. 1–3). IEEE.
[20] Tayal, D. K., & Yadav, S. K. (2016, March). Fast retrieval approach of sentimental analysis with implementation of bloom filter on Hadoop. In Computational Techniques in Information and Communication Technologies (ICCTICT), 2016 International Conference on (pp. 14–18). IEEE.
[21] Zhang, L., Bao, S., Guo, H., Zhu, H., Zhang, X., Cai, K., ... & Su, Z. (2010). EagleEye: Entitycentric business intelligence for smarter decisions. *IBM Journal of Research and Development*, 54(6), 1–1.
[22] Ahmed, S., & Danti, A. (2016). Effective sentimental analysis and opinion mining of web reviews using rule based classifiers. In *Computational Intelligence in Data Mining*—Volume 1 (pp. 171–179). Springer, New Delhi.
[23] Ahmed, S., & Danti, A. (2015, December). A novel approach for Sentimental Analysis and Opinion Mining based on SentiWordNet using web data. In Trends in Automation, Communications and Computing Technology (I-TACT-15), 2015 International Conference on (Vol. 1, pp. 1–5). IEEE.
[24] Dehkharghani, R., Mercan, H., Javeed, A., & Saygin, Y. (2014). Sentimental causal rule discovery from Twitter. *Expert Systems with Applications*, 41(10), 4950–4958.
[25] Balahur, A., & Perea-Ortega, J. M. (2015). Sentiment analysis system adaptation for multilingual processing: The case of tweets. *Information Processing & Management*, 51(4), 547–556.
[26] Meire, M., Ballings, M., & Van den Poel, D. (2016). The added value of auxiliary data in sentiment analysis of Facebook posts. *Decision Support Systems*, 89, 98–112.
[27] Cambria, E. (2016). Affective computing and sentiment analysis. *IEEE Intelligent Systems*, 31(2), 102–107.
[28] Pandey, A. C., Rajpoot, D. S., & Saraswat, M. (2017). Twitter sentiment analysis using hybrid cuckoo search method. *Information Processing & Management*, 53(4), 764–779.
[29] Blitzer, J., Dredze, M., & Pereira, F. (2007, June). Biographies, bollywood, boomboxes and blenders: Domain adaptation for sentiment classification. In ACL (Vol. 7, pp. 440–447).
[30] "Intelligent Recognition System for Identifying Items and Pilgrims", *NED University Journal of Research – Thematic Issue on Advances in Image and Video Processing*, ISSN: 2304-716X, pp. 17–23, May 2018.
[31] "Improving Hajj and Umrah Services Utilizing Exploratory Data Visualization Techniques", Hajj Forum 2016 - the 16th scientific Hajj research Forum, Organized by

the Custodian of the Two Holy Mosques Institute for Hajj Research, Umm Al-Qura University – King Abdulaziz Historical Hall, Makkah,Saudi Arabia, 24–25 May 2016.

[32] "Microscopic modeling of large-scale pedestrian-vehicle conflicts in the city of Madinah, Saudi Arabia", *Journal of Advanced Transportation*, DOI: 10.1002/atr.1201, Article first published online: 27 JUL 2012.

[33] "Trialing a Smart Face-recognition Computer System to Recognize Lost People Visiting the Two Holy Mosques", *Arab Journal of Forensic Sciences & Forensic Medicine (AJFSFM)*, ISSN: 1658–6786, Vol. 1, No. 8, Published by Naif Arab University for Security Sciences (NAUSS), 2018.

[34] "Exploratory Data Visualization for Smart Systems", Smart Cities 2015 – 3rd Annual Digital Grids and Smart Cities Workshop, Burj Rafal Hotel Kempinski, Riyadh, Saudi Arabia, May 2015.

[35] "Data Visualization to Explore Improving Decision-Making within Hajj Services", *Scientific Modelling and Research*, Vol. 2, No. 1, Pages: 9–18, DOI: 10.20448/808.2. 1.9.18, 1 June 2017. – "Velocity-Based Modeling of Physical Interactions in Dense Crowds", The Visual Computer, Springer, 3 June 2014.

[36] "Users' Evaluation of Rail Systems in Mass Events: Case Study in Mecca, Saudi Arabia", *Transportation Research Record: Journal of the Transportation Research Board*, Issue 2350, Pages 111–118, December 2013.

[37] "Right of Way: Asymmetric Agent Interactions in Crowds", *The Visual Computer: International Journal of Computer Graphics*, Vol. 29, No. 12, Pages: 1277–1292, Springer- Verlag Berlin Heidelberg, 2013, DOI: 10.1007/s00371-012-0769-x, Published online: December 2012.

[38] "Cybercrime on Transportation Airline", *Journal of Forensic Research*, ISSN: 2157–7145, Vol. 10, No. 4, 19 November 2019.

[39] "Protecting Medical Records against Cybercrimes within Hajj Period by 3-layer Security", *Recent Trends in Information Technology and Its Application*, Vol. 2, No. 3, Pages: 1–21, DOI: 10.5281/zenodo.3543455, 15 November 2019.

[40] Brendon et al., Geospatial Event Detection by Grouping Emotion Contagion in Social Media, *IEEE* (2020), pp. 159–170, volume- 6.

[41] Guoshuai et al., CAPER: Context-Aware Personalized Emoji Recommendation, *IEEE* (2020), pp. 1–13.

[42] Akshi et al., Hybrid context enriched deep learning model for fine-grained sentiment analysis in textual and visual semiotic modality social data, ELSEVIER (2020), DOI: 10.1016/j.ipm.2019.102141

[43] Asim et al., Adapting Deep Learning for Sentiment Classification of Code-Switched Informal Short Text, ACM(2020). [44] Abdullatif et al., Arabic Sentiment Analysis: A System.

[44] Rajput, P. K., Nagpal, G. and Aarti (2013) Feature Weighted Unsupervised Classification Algorithm and Adaptation for Software Cost Estimation. *International Journal of Computational Intelligence Studies. ISSN:* 1755–4977, 3(1), pp.74–93, 2013.

[45] Aarti, Sikka, G. and Dhir, R. (2016) An investigation on the effect of cross project data for prediction accuracy. *International Journal of System Assurance Engineering and Management, Springer*, 7(1), 1–26.

[46] Aarti, Sikka, G. and Dhir, R. (2016) An investigation on the metric threshold for fault-proneness. *International Journal of Education and Management Engineering (IJEME), MECS*, 1(1), 1–8.

[47] Aarti, Sikka, G. and Dhir, R. (2016) Identification of error prone classes and quality estimation using semi-supervised learning mechanism with limited fault data. In Proceeding of Recent Research in Mechanical, Electrical, Electronics, Civil, Computer Science and Information Technology (MECIT-2017), New Delhi.

[48] Aarti, Sikka, G. and Dhir, R. (2017) Threshold based empirical validation of object-oriented metrics on different severity levels. *International Journal of Intelligent Engineering Informatics (IJIEI), Inderscience*, 7(2/3), 231–262.

[49] Aarti, Sikka, G. and Dhir, R. (2018) Grey relational classification algorithm for software fault- proneness with SOM clustering. *International Journal of Data Mining, Modelling and Management (IJDMMM), Inderscience*, 12(1), 28–64.

[50] Aarti, Sikka, G. and Dhir, (2020) Novel Grey Relational Feature Extraction Algorithm for Software Fault-Proneness Using BBO (B-GRA). *Arabian Journal for Science and Engineering*, 45(4), 2645–2662.

[51] Aarti, Sikka, G. and Dhir, (2020), Hybrid semi-supervised SOM based clustered approach with genetic algorithm for software fault classification, *Feature Extraction and Classification Techniques for Text Recognition(book)*, SBN13: 9781799824060

Index

abnormal, 205, 210
abruptly, 28
abundantly, 110
accelerated, 45, 153
accessibility, 152
accidental, 219
accommodated, 153
accountability, 184
accumulation, 212–14
acquainted, 153
acquisition, 155, 156, 159, 160, 244
activities, 106, 156, 197
adaptation, 124, 225, 245, 246
additional, 73, 101, 155, 167, 168, 229
adopted, 5, 26, 114, 233
adsorbing, 22, 90
advantageous, 27
advertisements, 237
affinity, 58
agriculture, 73, 146
ailment, 73, 80
alcohol, 219
Alexander, 205, 222
all-around-based, 92
alongside, 32, 208
aluminium, 21
ambi-polar, 2
ammonia, 22, 90, 99, 102
analyte, 72, 92, 113
analytically, 60, 229
analyzes, 195
analyzing, 16, 34, 148, 155, 215, 219, 228, 234
angular, 210
annealing, 2
anomaly, 155, 159, 224
applying, 53, 73, 142, 210, 212, 237
appropriately, 27
assuming, 172
attenuation, 162, 171
avoided, 133, 197

back-end, 169
backward, 153
bandpass, 149
band-to-band, 71, 114
bandwidth, 156, 165, 196
baseline, 172
batteries, 22
behind, 79, 85, 91, 93, 199, 228
Bessel, 59, 60

biassing, 16
BIG-DATA, 150
Bioelectronics, 67, 68, 119
bio-inspired, 149, 159, 223
bio-molecules, 3, 18, 109
bioreceptors, 113
bipolar, 148, 159
Boltzmann's, 47, 52, 56, 58
bootstrap, 172
boundary, 16, 54, 111
bridged, 32, 34
Burke, 25

calculating, 25, 168
calibrated, 7, 45, 74
cancellation, 213
capture, 22
careful, 118
categorization, 154
centralized, 162, 167, 168
character-based, 243
chemicals, 3
chip, 40, 74, 148, 151
classifications, 199
coefficient, 140, 141
collusion, 169
communicate, 163, 165, 169, 229
comparision, 88, 91
comprehend, 25, 166
concave, 82
conductivities, 155
connecting, 187, 191
constituents, 22
contributions, 148
converting, 209
corpus-based, 230
coulomb, 34, 35
creating, 70, 228, 237
crystal, 49, 50, 57

data-centric, 175, 200
decade, 2, 4, 106, 205–7
decimal, 212
deduced, 20–22, 24, 32
degree, 236
demonstrated, 28, 40, 48, 125, 126, 220
deoxyribonucleic, 4, 126
descendent, 130
details, 124, 144, 173
developments, 151

249

dictionaries, 230
difficulty, 72
diffuse, 64
digitizer, 7
disorders, 155, 156, 218
distinction, 131
divulged, 188
DM-FETs, 70
document-level, 230, 234, 244
donor, 51, 52, 54, 107
downscaling, 41, 42
drift-diffusion, 111

ecological models, 139
effectiveness, 155, 208
efficiency, 4, 80, 133, 154, 156, 191, 208, 214, 220, 222, 223, 231, 233–38, 244
electromyographic, 207
electrophysiological, 64
emission, 71, 106
emphasized, 170, 236
encapsulated, 21, 22
endohedral-position, 20
engineered, 83
entity-level, 234
equations, 44, 47, 50, 56, 60, 61, 67, 111, 214
equipment, 96, 113, 151, 154
evolutionary, 124, 177, 221, 233
exchange-correlation, 25
extraordinary, 22

fabricated, 20, 40, 47, 74, 156
facing, 188
failure, 150, 167, 219
Federal, 163
fetched, 212
fields, 4, 67, 72, 75, 97, 151, 170, 188, 210, 221
filling, 3, 114, 136
filters, 149, 222
finite, 210
Flowchart, 128
f-measure, 232
focuses, 5, 11
formatting, 133
forwarded, 169, 172
fuel, 162
functioning, 3, 114, 115, 155, 198
fusion, 134, 223, 224

gain, 153, 155, 156, 159, 171, 220, 244
gained, 20, 210
gallium, 44
gaps, 3, 161, 166, 167, 170, 192
gate-dielectric, 94
gate-metal, 88
gate-source, 106
gating, 77, 113

geometrical, 16, 44, 57, 58, 60
globe, 152
gradient-based, 124

halt, 139
handle, 134, 183, 184
harmful, 205
heavily, 4, 183
Hetero-structures, 76
Hexadecimal, 130
holistic, 198
hydrogen, 20, 22, 36, 99, 102
hysteresis, 110

ideal, 29, 35, 204
identical, 137
identification, 3, 181, 186, 188, 225, 246
immediately, 138
immobilized, 5, 9, 10, 79, 110, 117
implementing, 106, 146, 148, 154, 205, 208, 210, 215, 221, 222
implies, 27, 29, 206
incident, 93, 94, 177
inclusion, 133, 156
index, 210
Inequality, 140
infrared, 21, 93, 95, 103
Innovation, 147, 151
instrumentation, 156
interacting, 174
interference, 74, 157
internet-based, 171
intrusion, 170, 178, 200, 201

jamming, 165, 176, 182, 191, 192, 201
JLFET, 4
junction-less, 91, 101

keeping, 70, 131, 151, 166, 171, 175, 215
K-map, 214, 215
k-means, 233
knowledge-based, 228
Kramers-Brillouin, 75
k-values, 82, 85
k-vector, 140

label, 3, 10, 71, 72, 97, 102, 121, 229
labeled, 45, 46, 57, 110, 134, 204, 228
large-scale, 148, 152, 184, 198, 228, 244, 246
leakage, 2, 16, 17, 42, 86, 93, 113, 120, 121, 153
lifestyle, 162
limits, 17, 40, 71, 72
lithium, 21
low-consumption, 118

marginally, 243
matching, 6, 70, 136

Index   251

membrane, 155
metric, 77, 83, 93, 101, 115, 214, 217, 220, 246
metrices, 214
microscope, 27
moderate, 16
Mole, 116
monitor, 72, 151, 156, 221, 225
MOSFET-based, 79, 88, 93, 96
multiply-accumulate, 207, 212
multipoint, 128
multi-purpose, 149
multi-sensor, 149

nano, 3, 18, 35, 36, 66, 68, 98, 99, 101, 119
nano-carrier, 22
nanocavity, 81, 110, 112
nanocluster, 21
nano-sensor, 70, 96
nano-sensors, 70, 71, 75
n-channel, 107
near-infrared, 94, 103
network–based, 183, 203, 211
next-generation, 96
Non-emergency, 162
non-linearity, 32
not-for-profit-sectors, 35
numerical, 231

objectivity, 172
occupancy, 51
offset, 4
one-dimensional, 40, 42, 55, 66, 111
operand, 210, 214
operate, 6
optimizes, 155
optoelectronic, 70, 97
opto-electronic, 21
outperform, 208
overcoming, 4, 101, 233
overload, 192
overpowering, 190

package, 231
packet, 163, 191, 192, 200
participate, 32, 162, 165, 167, 171, 184
participating, 27, 162, 164, 165, 170
pathfinding, 125
penetrating, 60
periodic, 29, 49
personality, 155
photogeneration, 94
photonic, 2, 18
p-n-p-n, 79
pointers, 133
potassium, 21
predictor, 149, 205, 235

presence, 21, 43, 70, 72, 74, 79, 98, 114, 115, 125
producing, 130, 133, 207
promotes, 184

quantify, 74
quantity, 73, 85, 90, 92, 174
quantization, 43
quasi-ballistic, 48, 67
quasi-Fermi, 54
quasi-synchronous, 148
quick-response, 96

radiology, 207
radix, 148
random-dopant, 95
react, 85, 91
reading, 228
received, 165, 169, 188, 191, 195–97
reception, 196
reconstruct, 131
records, 237, 238, 246
refilling, 3
reform, 124
relate, 184
relatively, 153
remind, 156
represent, 13, 24, 26, 29, 130, 208
resistance, 21, 41, 48, 74, 121
reunited, 170
riffling, 129
robust, 20, 97, 124, 125, 131, 143, 166, 167
routed, 163
rule-based, 233

sample, 72, 133, 155, 214
satisfaction, 156
satisfying, 149
scaled, 4, 41, 44, 66, 106
Schrödinger, 47, 49
scientist, 71, 145
segment, 169
Sekanina, 215, 223
semiconducting, 43, 98, 119
semiconductors, 44, 45, 60, 67, 68
serializer, 156
shielding, 156
short-gate, 80, 96, 101
significance, 238
single-chip, 72, 77
single-crystal, 48
Single-Point, 134
sodium-ion, 22
spin-electronic, 21
Substractions, 150
symmetrically, 107
synthesis, 20

system-on-chip, 205

tcavity, 7, 12, 13, 15, 16
temporal, 208
terminal, 41, 42, 46
thallium, 21
theoretical, 43, 120, 178
time-dependent, 49
Time-division, 200
tolerant, 168, 182, 199
trained, 205, 220, 221, 231, 236, 244
trigram, 235, 236
tunability, 113
two-parent, 128, 136

ultra-small, 95
ultrasound, 148
ultra-thin, 42
unbiased, 230, 233, 234
underweight, 219
uniformly, 114
unreliably, 164, 167, 173
unsupervised, 204, 246
unwanted, 193, 237, 238

U-shape, 77, 100
utilization, 21, 24, 147, 150, 152–54, 244

valley, 58
valuable, 124
varied, 109, 199
Vehicle-to-Vehicle, 200
Very-large-scale, 151

wafer, 117
wafers, 117
wall, 207
warnings, 162, 199
waves, 197, 210
well-researched, 74
widespread, 70, 205
worldwide, 73, 187
worn, 210
WorNet, 231

zero, 133, 135, 149, 166
zero-day, 183
zeta-polarized, 25
zone, 51, 74